Farah Kanbar

Asymptotic and Stationary Preserving Schemes for Kinetic and Hyperbolic Partial Differential Equations

Farah Kanbar

Asymptotic and Stationary Preserving Schemes for Kinetic and Hyperbolic Partial Differential Equations

Würzburg
University Press

Dissertation, Julius-Maximilians-Universität Würzburg
Fakultät für Mathematik und Informatik, 2022
Gutachter: Prof. Dr. Christian Klingenberg, Prof. Dr. Rony Touma

Impressum

Julius-Maximilians-Universität Würzburg
Würzburg University Press
Universitätsbibliothek Würzburg
Am Hubland
D-97074 Würzburg
www.wup.uni-wuerzburg.de

© 2023 Würzburg University Press
Print on Demand

Coverdesign: Holger Schilling

ISBN 978-3-95826-210-2 (print)
ISBN 978-3-95826-211-9 (online)
DOI 10.25972/WUP-978-3-95826-211-9
URN urn:nbn:de:bvb:20-opus-301903

Acknowledgments

I would like to start by expressing my gratitude to my supervisor Prof. Dr. Christian Klingenberg for his unlimited support on a professional and personal level.

I would like to extend my sincere thanks to my co-supervisor Prof. Dr. Rony Touma for always being supportive, ever since I started my research career.

I would also like to thank our collaborator Dr. Min Tang for the valuable scientific discussions and the thorough insights.

I thank my colleagues throughout these four years for their insightful comments, suggestions, support, and patience.

My deep feelings go all the way to Lebanon, to my family and fiancé whom I am away from. Thank you for your unwavering support and belief in me.

A big thanks goes to my international friends in Germany who have made this journey enjoyable.

Preface

This PhD thesis arose from a research collaboration between my work group at Würzburg University and Prof. Fritz Röpke, an astrophysicist in Heidelberg. He studies numerically the temporal evolution of stars. His need to numerically solve the compressible Euler equations with gravity near a stationary solution gave rise to the questions addressed in this PhD thesis. The research itself was done in collaboration both with Prof. Rony Touma in Lebanon and also with Prof. Min Tang from Shanghai, China.

This dissertation treats an important topic in applied mathematics. Numerical methods for hyperbolic balance laws of fluid type have typically been devised for the one-space dimensional case, for example the widely used Godunov-type methods. In applications flows need to be computed in three space dimension. It has been shown in the PhD thesis of Wasilij Barsukow that three important flow features in three space dimension are: asymptotic limits, stationary solutions and vortices. The numerical methods should be devised in such a way that the numerical discretization preserves these important multi dimensional features. One dimensional methods have difficulties preserving these features. This thesis by Farah Kanbar goes some way towards the goal of devising schemes for non-linear partial differential equations that preserve these features.

This thesis focuses on two of the above features, namely stationary solutions and asymptotic limits:

- stationary solutions are solutions of the hyperbolic balance law that do not change in time. The discretization should be such, that it exactly maintains a particular numerical discretization of stationary solutions.

- asymptotic limits refers to hyperbolic balance laws which are endowed with a small parameter. Taking the limit of this parameter to zero, one obtains a new partial differential equation. The example coming for the astrophysical application, mentioned in the beginning, is the limit of compressible flow equations to incompressible flow. Another limit is the limit of mesoscopic kinetic equations to their macroscopic counterpart. The numerical scheme should be devised in such a way, that the discretization of the balance law with a small parameter converges to a discretization of the limit equation as the small parameter goes to zero.

The overall outcome of this thesis is that one needs an interplay between stationary preserving schemes and asymptotic preserving schemes, in order to achieve numerical schemes that are able to preserve both features. In the literature these are examples of so called structure preserving schemes. They play an important role in applications.

The success of the research in this PhD thesis can be gauged by the five research publications in international journals that came out of this.

Thus this PhD thesis contributed in a significant way towards the goals of the questions in numerical analysis raised by Fritz Röpke.

Würzburg, in Feb. 2023
Christian Klingenberg

Abstract

In this thesis, we are interested in numerically preserving stationary solutions of balance laws. We start by developing finite volume well-balanced schemes for the system of Euler equations and the system of Magnetohydrodynamics (MHD) equations with gravitational source term. Since fluid models and kinetic models are related, this lead us to investigate Asymptotic Preserving (AP) schemes for kinetic equations and their ability to preserve stationary solutions. Kinetic models typically have a stiff term, thus AP schemes are needed to capture good solutions of the model. For such kinetic models, equilibrium solutions are reached after large time. Thus we need a new technique to numerically preserve stationary solutions for AP schemes. We find a criterion for Stationary Preserving (SP) schemes for kinetic equations which states, that AP schemes under a particular discretization are also SP. In an attempt to mimic our result for kinetic equations in the context of fluid models, for the isentropic Euler equations we developed an AP scheme in the limit of the Mach number going to zero. Our AP scheme is proven to have a SP property under the condition that the pressure is a function of the density and the latter is obtained as a solution of an elliptic equation. The properties of the schemes we developed and its criteria are validated numerically by various test cases from the literature.

Zusammenfassung

In dieser Arbeit interessieren wir uns für numerisch erhaltende stationäre Lösungen von Erhaltungsgleichungen. Wir beginnen mit der Entwicklung von well-balanced Finite-Volumen Verfahren für das System der Euler-Gleichungen und das System der MHD-Gleichungen mit Gravitationsquell term. Da Strömungsmodelle und kinetische Modelle miteinander verwandt sind, untersuchen wir asymptotisch erhaltende (AP) Verfahren für kinetische Gleichungen und ihre Fähigkeit, stationäre Lösungen zu erhalten. Kinetische Modelle haben typischerweise einen steifen Term, so dass AP Verfahren erforderlich sind, um gute Lösungen des Modells zu erhalten. Bei solchen kinetischen Modellen werden Gleichgewichtslösungen erst nach langer Zeit erreicht. Daher benötigen wir eine neue Technik, um stationäre Lösungen für AP Verfahren numerisch zu erhalten. Wir finden ein Kriterium für stationär-erhaltende (SP) Verfahren für kinetische Gleichungen, das besagt, dass AP Verfahren unter einer bestimmten Diskretisierung auch SP sind. In dem Versuch unser Ergebnis für kinetische Gleichungen im Kontext von Strömungsmodellen nachzuahmen, haben wir für die isentropen Euler-Gleichungen ein AP Verfahren für den Grenzwert der Mach-Zahl gegen Null, entwickelt. Unser AP Verfahren hat nachweislich eine SP Eigenschaft unter der Bedingung, dass der Druck eine Funktion der Dichte ist und letztere als Lösung einer elliptischen Gleichung erhalten wird. Die Eigenschaften des von uns entwickelten und seine Kriterien werden anhand verschiedener Testfälle aus der Literatur numerisch validiert.

Abbreviations

AP Asymptotic Preserving

CFL Courant–Friedrichs–Lewy

CTM Constrained Transport Method

FV Finite Volume

IMEX Implicit-Explicit

MAC Marker and Cell

MHD Magnetohydrodynamics

NT Nessyahu-Tadmor

PDE Partial Differential Equations

SP Stationary Preserving

TVD Total Variation Diminishing

UC Unstaggered Central

UGKS Unified Gas Kinetic Scheme

1D one-dimensional

2D two-dimensional

Contents

Chapter 1

Introduction

The models

Partial Differential Equations (PDE): A partial differential equation is an equation that imposes relations between partial derivatives of a function of more than one variable. The function is the unknown to be found. Partial differential equations are largely used in applied mathematics, physics and engineering. The equations play a big role in the modern scientific understanding of sound, heat, diffusion, electrostatics, electrodynamics, thermodynamics, fluid dynamics, elasticity, general relativity, and quantum mechanics, etc. There are three types of partial differential equations: hyperbolic, parabolic and elliptic. In this thesis we focus on hyperbolic partial differential equations. The solutions of hyperbolic equations are "wave-like", such that perturbations of the initial or the boundary data travel along the characteristics of the equation.

Fluid Mechanics: Fluid Mechanics is a division of physics concerned with the mechanics of the fluid under internal and external forces. It studies fluids in their static or dynamic states. Fluid dynamics is a subsection of fluid mechanics that decscribes the flow of fluids (liquids and gases) and it is divided into two other subsections: aerodynamics, the study of air and other gases in motion, and hydrodynamics, the study of liquids in motion. The solution to a fluid dynamics problem typically involves the calculation of various physical properties of the fluid, such as flow velocity, pressure, density, and temperature, as functions of space and time. In this thesis three fluid models are considered. The first model is the system of Euler equations with gravitational source term which we will introduce in chapter 2. This system is widely studied because of its importance in modelling physical phenomena such as astrophysical and atmospheric phenomena including supernova explosions [56], climate modelling, and weather forecasting [12]. A special case of the Euler equations are the isentropic Euler equations which we will also see in chapter 4. The system of MHD equations, defined in chapter 2, is a combination of the Euler equations of fluid dynamics and Maxwell's equations of electromagnetism. A gravitational source term is added to the ideal MHD equations in this work.

Kinetic theory of gases: In chapter 3, several kinetic models are considered. They describe a gas as a large number of identical submicroscopic particles (atoms or molecules), all of which are in constant, rapid, and random motion. Their size is assumed to be much smaller than the average distance between the particles. Kinetic models describe the time evolution of probability density distribution of particles that travel freely for a certain distance, and then change their directions due to collision or scattering. They usually include a transport term that takes into account the movement of the particles, and integral terms that take into account the scattering, tumbling or colliding.

Numerical Methods

Solving partial differential equations is a broad topic in applied mathematics. However, finding exact solutions for these equations is not always possible. There is, correspondingly, a vast amount of modern mathematical and scientific research on methods to numerically approximate solutions of certain partial differential equations using computers. A numerical method for partial differential equations is a mathematical tool designed to find numerical solutions for the equation. The implementation of a numerical method with an appropriate convergence check in a programming language is called a numerical algorithm. Computing a numerical solution is finding the discrete version of the continuous solution of the PDE via a numerical algorithm.

Finite Volume (FV) Central Scheme: To design a numerical scheme, one has to consider time and space. A finite volume method is a reformulation of Godunov's method for the spatial discretization and is based on averaging the conserved variables in each cell and approximating the fluxes between the cells. We use finite volume central schemes as base scheme in chapter 2 which relies on the fact that central schemes are easy to implement and robust finite volume schemes that avoid the time consuming process of solving Riemann problems arising at the cell interfaces. Furthermore, central schemes have proven to be efficient schemes for the simulation of systems of hyperbolic conservation laws. Nessyahu and Tadmor [61] have introduced the Nessyahu-Tadmor (NT) scheme, a non-oscillatory central finite volume scheme that is based on evolving piecewise linear numerical solution on two staggered grids. Useful extensions of the NT scheme to multi-space dimensions followed in [5, 36, 46, 6, 7, 44, 77]. These extensions were successfully used to solve problems arising in aerodynamics, hydrodynamics, and magnetohydrodynamics [7, 43, 74, 76].

In order to avoid switching between an original and a staggered grid in the NT-type schemes, Unstaggered Central (UC) schemes for hyperbolic systems of conservation laws were developed in [45, 73], where the numerical solution is evolved on a single grid. The UC schemes were then extended to hyperbolic balance laws such as shallow water equations on variable waterbeds, Ripa systems, and Euler with gravity systems [80, 78, 79, 74]. The main goal of the UC schemes is to evolve the numerical solution on a single grid and to use a staggered ghost grid in an intermediate step, followed by a back projection step, see figure 2.3.

Schemes for Kinetic Models: Three different AP schemes for three different kinetic models are considered in chapter 3. Developing the three AP schemes is not a focus of this thesis as they are taken from the literature. However, their SP property and whether they satisfy the proposed criterion or not are evaluated in this thesis. The three schemes are parity equations-based scheme for the neutron transport equation, Unified Gas Kinetic Scheme (UGKS) for the chemotaxis kinetic model, and IMEX scheme with the Penalization method for the Boltzmann equation.

Marker and Cell (MAC) Schemes: A finite difference staggered approached, suggested by Goudon et al. [34] is chosen in chapter 4. The staggered discretization follows the principles of MAC schemes [38]. The idea of MAC is to place the variables of the system in different locations on the grid. The detailed description of the method can be found in chapter 4.

Properties of the Numerical Methods

Well-balanced Schemes: Of particular interest are stationary solutions of the PDE. Those solutions need to be taken into account in the discretization of the scheme. We define well-balanced schemes as schemes that are designed to preserve a prior known stationary solution. One example of these solutions is the case of zero velocity called hydrostatic equilibrium. One way to fulfil the well-balanced requirement of the numerical scheme is by designing the discretization in the source term in the balance law by following that of the divergence of the flux function. There are several methods to develop a well-balanced scheme that all require that the steady state is known or given. Several attempts were previously made for designing well-balanced schemes for balance laws [8, 80, 20, 22, 84, 85, 86, 35, 81, 68, 10, 9, 23, 24, 53, 18, 82, 19].

Asymptotic Preserving Schemes: The parameter ε which is the Knudsen number (for kinetic models), is the ratio of the mean free path and the domain typical length scale [26]. This parameter pops up in the equations after rescaling, creating a stiff term where it is located. A similar parameter for the fluid models (Mach number) also appears in the equations after rescaling, leaving stiff terms behind. Numerical schemes do not behave well when such parameter exists. This is because when ε goes to zero it causes very small time steps. Hence, AP schemes that allow very small values of this parameter become popular in this area. A numerical scheme is AP if when the parameter goes to zero in the discretized scheme, it converges to a good discretization of the corresponding limit model. The aim of AP schemes is to discretize the stiff term of the equation implicitly, which leads to an Implicit-Explicit (IMEX) discretization of the model. The main advantage of AP schemes is that their stability and convergence are independent of the parameter.

Stationary Preserving Schemes For schemes such as AP schemes, the solution after some time reaches a quasi-stationary state, meaning numerically that the difference between the global equilibrium and the solution after finite time is smaller than machine precision. Which means the steady solution is not given and is not known. For this reason, more than well-balancing, we need a discretization that preserves any state that might show up as time evolves. Thus, it is of interest to have a numerical scheme that maintains stationary solutions up to machine precision. We call such schemes SP schemes. A scheme is SP if the following two requirements are satisfied:

- The discrete stationary solution provides a good approximation for the steady state solution;

- Starting from a discrete stationary solution, the solution of the time evolutionary problem does not change.

Numerically, one can test that the time evolutionary problem converges to a discrete stationary solution after finite time, and their difference is smaller than machine precision.

Organization of the Thesis

Chapter 1 provides a background for the topics covered in this thesis with a review of prior works. In chapter 2, we present one-dimensional (1D) and two-dimensional (2D) well-balanced central schemes with applications to the Euler and MHD equations with gravitational source term. Then we present three schemes for kinetic models in chapter 3. The three schemes are proven to satisfy a common criterion. In chapter 4, an AP scheme for the isentropic Euler equations with gravitational source term is developed and then proven under certain conditions to be SP. And finally, we conclude by proposing some future work.

Chapter 2

Well-balanced Central Schemes with the Subtraction Method

2.1 Introduction

As mentioned in the introduction, the first task in my project was to develop a well-balanced, unstaggered, second-order, finite volume central scheme for the Euler equations with gravitational source term via a subtraction method [51]. A normal extension was to apply the obtained scheme to the system of MHD equations with gravitational source term [52]. The developed numerical schemes avoid solving Riemann problems at the cell interfaces and avoid switching between an original and a staggered grid. Their main feature is that they are capable of preserving any given steady state up to machine accuracy by updating the numerical solution in terms of a relevant reference solution. The methodology proposed results in a well-balanced scheme capable of capturing any given steady state. In this work we follow a special reconstruction in the conservative variables that will fulfil the well-balanced requirement and will allow a proper capture of the steady states. This well-balanced approach will be blended with the unstaggered central finite volume scheme for hyperbolic systems of conservation laws [73]. The proposed method follows the subtraction method developed by Berberich, Chandrashekar and Klingenberg [8]. It consists of evolving the error function between the vector of conserved variables and a given steady state, instead of evolving the vector of conserved variables. Our scheme is then implemented and used to solve classical problems from the recent literature. We consider the Courant–Friedrichs–Lewy (CFL) convergence condition for our numerical scheme. It enforces an upper bound on the time step, otherwise the explicit scheme produces irrelevant results. In sections 2.2 and 2.3, we present the 1D and 2D schemes for general balance laws respectively. The discretization is proven to be Total Variation Diminishing (TVD) in section 2.4. We apply the developed schemes to the 1D and 2D Euler and then to the 2D MHD system in section 2.5.

2.2 1D Unstaggered Well-balanced FV Central Scheme

In this section we develop a new 1D unstaggered well-balanced central scheme for balance laws. The proposed method follows the subtraction method introduced in

[8]. Consider the 1D balance law given by

$$\begin{cases} \mathbf{u}_t + f(\mathbf{u})_x = S(\mathbf{u}, x), & x \in \Omega \subset \mathbb{R}, \ t > 0 \\ \mathbf{u}(x, 0) = \mathbf{u}_0(x) \end{cases} \tag{2.1}$$

where $\mathbf{u}$ is the vector of conserved variables, $f(\mathbf{u})$ is the flux function and $S(\mathbf{u}, x)$ is the source term. We consider for our computational domain Ω an interval of the real axis, and we partition it using the control cells defined to be the subintervals $C_i = \left[x_{i-\frac{1}{2}}, x_{i+\frac{1}{2}} \right]$ of equal width $\Delta x = x_{i+\frac{1}{2}} - x_{i-\frac{1}{2}}$ and centered at the nodes x_i. We also define the dual ghost cells $D_{i+\frac{1}{2}} = [x_i, x_{i+1}]$ with centers $x_{i+\frac{1}{2}} = x_i + \frac{\Delta x}{2}$. The main and the staggered 1D grids are illustrated in figure (2.1). The time-step will be denoted by Δt, and it is computed using the CFL condition,

$$\Delta t = \text{CFL} \frac{\Delta x}{\max(|\lambda_k|)}, \tag{2.2}$$

where $0 \leq \text{CFL} \leq 0.5$ and λ_k is the maximum eigenvalue of the flux jacobian matrix $\frac{\partial f(\mathbf{u})}{\partial \mathbf{u}}$. For a positive integer n we set $t^{n+1} = t^n + \Delta t$.

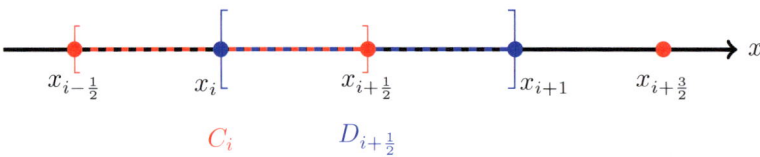

Figure 2.1: The 1D grid partitioned into control cells $C_i = \left[x_{i-\frac{1}{2}}, x_{i+\frac{1}{2}} \right]$ and dual cells $D_{i+\frac{1}{2}} = [x_i, x_{i+1}]$.

We assume that the numerical solution $\mathbf{u}_i^n$ at time t^n is known at the nodes x_i where $\mathbf{u}_i^n$ is used to approximate the exact solution $\mathbf{u}(x_i, t^n)$. We start the derivation of our numerical scheme by first assuming that $\tilde{\mathbf{u}}$ is a given stationary solution of system (2.1), and we follow the subtraction method introduced in [8]. Let $\Delta \mathbf{u} = \mathbf{u} - \tilde{\mathbf{u}}$, we substitute $\mathbf{u} = \Delta \mathbf{u} + \tilde{\mathbf{u}}$ in the balance law in system (2.1),

$$(\Delta \mathbf{u} + \tilde{\mathbf{u}})_t + f(\Delta \mathbf{u} + \tilde{\mathbf{u}})_x = S(\Delta \mathbf{u} + \tilde{\mathbf{u}}, x), \tag{2.3}$$

and taking into account that $\tilde{\mathbf{u}}$ is a stationary solution, this results in,

$$(\Delta \mathbf{u})_t + f(\Delta \mathbf{u} + \tilde{\mathbf{u}})_x = S(\Delta \mathbf{u} + \tilde{\mathbf{u}}, x). \tag{2.4}$$

On the other hand, since $\tilde{\mathbf{u}}$ is a stationary solution of (2.1), then the balance law reduces to,

$$f(\tilde{\mathbf{u}})_x = S(\tilde{\mathbf{u}}, x) \tag{2.5}$$

Subtracting (2.5) from (2.4) leads to,

$$(\Delta \mathbf{u})_t + [f(\Delta \mathbf{u} + \tilde{\mathbf{u}}) - f(\tilde{\mathbf{u}})]_x = S(\Delta \mathbf{u} + \tilde{\mathbf{u}}, x) - S(\tilde{\mathbf{u}}, x). \tag{2.6}$$

But since $S(\mathbf{u}, x)$ is a linear functional in terms of the conserved variables, then equation (2.6) simplifies to,

$$(\Delta\mathbf{u})_t + [f(\Delta\mathbf{u} + \tilde{\mathbf{u}}) - f(\tilde{\mathbf{u}})]_x = S(\Delta\mathbf{u}, x). \tag{2.7}$$

Our proposed numerical scheme follows a classical finite volume construction; we define the piecewise linear interpolants that approximate the exact solution $\Delta\mathbf{u}(x, t^n)$ on the cells C_i as follows:

$$\mathcal{L}_i(x, t^n) = \Delta\mathbf{u}_i^n + (x - x_i)\frac{(\Delta\mathbf{u}_i^n)'}{\Delta x}, \qquad \forall x \in C_i \tag{2.8}$$

where $\frac{(\Delta\mathbf{u}_i^n)'}{\Delta x}$ is a limited numerical spatial derivative approximating $\frac{\partial\Delta\mathbf{u}}{\partial x}(x, t^n)|_{x=x_i}$ and the slope $(\Delta\mathbf{u}_i^n)'$ is obtained using the (MC-θ) limiter (2.9). The numerical base scheme evolves a piecewise linear solution $\mathcal{L}_i(x, t)$, in each cell C_i, that approximates the analytic solution $\Delta\mathbf{u}(x, t)$ with

$$\Delta\mathbf{u}_i^n = \frac{1}{\Delta x}\int_{C_i}\mathcal{L}_i(x, t^n)\,dx \approx \frac{1}{\Delta x}\int_{C_i}\Delta\mathbf{u}(x, t^n)\,dx.$$

Before proceeding with the presentation of the numerical scheme we introduce some notations that will be used throughout the remaining of the chapter. In order to approximate the spatial numerical derivatives, the (MC-θ) limiter is considered which is defined as

$$(\Delta\mathbf{u}_i^n)' = \text{minmod}\left[\theta\left(\Delta\mathbf{u}_i^n - \Delta\mathbf{u}_{i-1}^n\right), \frac{\Delta\mathbf{u}_{i+1}^n - \Delta\mathbf{u}_{i-1}^n}{2}, \theta\left(\Delta\mathbf{u}_{i+1}^n - \Delta\mathbf{u}_i^n\right)\right] \tag{2.9}$$

where θ is a parameter that takes any value $1 < \theta < 2$, while the minmod function is defined as:

$$\text{minmod}(a, b, c) = \begin{cases} \text{sign}(a)\min\{|a|, |b|, |c|\}, & \text{if sign}(a) = \text{sign}(b) = \text{sign}(c) \\ 0, & \text{Otherwise.} \end{cases}$$

Next, we integrate (2.7) over the domain $R_{i+\frac{1}{2}}^n = D_{i+\frac{1}{2}} \times [t^n, t^{n+1}]$:

$$\iint_{R_{i+\frac{1}{2}}^n}(\Delta\mathbf{u})_t + [f(\Delta\mathbf{u} + \tilde{\mathbf{u}}) - f(\tilde{\mathbf{u}})]_x\,dR = \iint_{R_{i+\frac{1}{2}}^n}S(\Delta\mathbf{u}, x)\,dR. \tag{2.10}$$

We apply Green's formula to the double integral on the left-hand side of equation (2.10), which allows us to change the double integral into a line integral by the following formula:

$$\iint_R\left(\frac{\partial Q}{\partial x} - \frac{\partial P}{\partial y}\right)dxdy = \oint_R(Pdx + Qdy),$$

7

with $\frac{\partial Q}{\partial x} = [f(\Delta \mathbf{u} + \tilde{\mathbf{u}}) - f(\tilde{\mathbf{u}})]_x$ and $\frac{\partial P}{\partial y} = -(\Delta \mathbf{u})_t$. Equation (2.10) writes as:

$$\oint_{\partial R^n_{i+1/2}} [f(\Delta \mathbf{u} + \tilde{\mathbf{u}}) - f(\tilde{\mathbf{u}})]dt - \Delta \mathbf{u}dx = \int_{t^n}^{t^{n+1}} \int_{x_i}^{x_{i+1}} S(\Delta \mathbf{u}, x)dxdt, \qquad (2.11)$$

where the boundary of the rectangle $R^n_{i+1/2}$ is $\partial R^n_{i+1/2} = [x_i, x_{i+1}] \cup [t^n, t^{n+1}] \cup [x_{i+1}, x_i] \cup [t^{n+1}, t^n]$ plotted in figure 2.2.

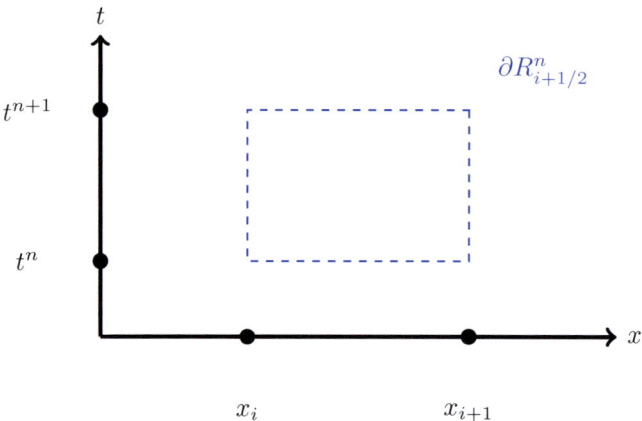

Figure 2.2: The boundary $\partial R^n_{i+1/2}$ (dashed) in the space-time plane.

Dividing the line integral over the four segments, we get:

$$\int_{x_i}^{x_{i+1}} \left[[f((\Delta \mathbf{u} + \tilde{\mathbf{u}})(x, t^n)) - f(\tilde{\mathbf{u}}(x, t^n))]dt - \Delta \mathbf{u}(x, t^n)dx \right]$$

$$+ \int_{t^n}^{t^{n+1}} \left[[f((\Delta \mathbf{u} + \tilde{\mathbf{u}})(x_{i+1}, t)) - f(\tilde{\mathbf{u}}(x_{i+1}, t))]dt - \Delta \mathbf{u}(x_{i+1}, t)dx \right]$$

$$+ \int_{x_{i+1}}^{x_i} \left[[f((\Delta \mathbf{u} + \tilde{\mathbf{u}})(x, t^{n+1})) - f(\tilde{\mathbf{u}}(x, t^{n+1}))]dt - \Delta \mathbf{u}(x, t^{n+1})dx \right]$$

$$+ \int_{t^{n+1}}^{t^n} \left[[f((\Delta \mathbf{u} + \tilde{\mathbf{u}})(x_i, t)) - f(\tilde{\mathbf{u}}(x_i, t))]dt - \Delta \mathbf{u}(x_i, t)dx \right]$$

$$= \int_{t^n}^{t^{n+1}} \int_{x_i}^{x_{i+1}} S(\Delta \mathbf{u}, x)dxdt \qquad (2.12)$$

Splitting the integrals and rearranging them simplifies equation (2.12) to:

$$-\int_{x_i}^{x_{i+1}} \Delta\mathbf{u}(x,t^n)dx + \int_{t^n}^{t^{n+1}} [f((\Delta\mathbf{u}+\tilde{\mathbf{u}})(x_{i+1},t)) - f(\tilde{\mathbf{u}}(x_{i+1},t))]dt$$

$$+\int_{x_i}^{x_{i+1}} \Delta\mathbf{u}(x,t^{n+1})dx - \int_{t^n}^{t^{n+1}} [f((\Delta\mathbf{u}+\tilde{\mathbf{u}})(x_i,t)) - f(\tilde{\mathbf{u}}(x_i,t))]dt$$

$$= \int_{t^n}^{t^{n+1}} \int_{x_i}^{x_{i+1}} S(\Delta\mathbf{u},x)dxdt. \quad (2.13)$$

The following integrals are approximated using second-order quadratures,

$$\int_{x_i}^{x_{i+1}} \Delta\mathbf{u}(x,t^n)dx = \Delta x \mathcal{L}_i(x_{i+\frac{1}{2}},t^n) = \Delta x \Delta\mathbf{u}_{i+\frac{1}{2}}^n,$$

and

$$\int_{x_i}^{x_{i+1}} \Delta\mathbf{u}(x,t^{n+1})dx = \Delta x \mathcal{L}_i(x_{i+\frac{1}{2}},t^{n+1}) = \Delta x \Delta\mathbf{u}_{i+\frac{1}{2}}^{n+1}.$$

Finally, the calculations on the left-hand side of equation (2.13) yield,

$$\Delta\mathbf{u}_{i+\frac{1}{2}}^{n+1} = \Delta\mathbf{u}_{i+\frac{1}{2}}^n - \frac{1}{\Delta x}\left[\int_{t^n}^{t^{n+1}} \{f((\Delta\mathbf{u}+\tilde{\mathbf{u}})(x_{i+1},t)) - f((\Delta\mathbf{u}+\tilde{\mathbf{u}})(x_i,t))\}\,dt\right]$$

$$+ \frac{\Delta t}{\Delta x}f(\tilde{\mathbf{u}}(x_{i+1})) - \frac{\Delta t}{\Delta x}f(\tilde{\mathbf{u}}(x_i)) + \frac{1}{\Delta x}\int_{t^n}^{t^{n+1}} \int_{x_i}^{x_{i+1}} S(\Delta\mathbf{u},x)dxdt. \quad (2.14)$$

The flux integrals in equation (2.14) are estimated using the midpoint quadrature rule as follows:

$$\int_{t^n}^{t^{n+1}} f((\Delta\mathbf{u}+\tilde{\mathbf{u}})(x_i,t))dt \approx f((\Delta\mathbf{u}+\tilde{\mathbf{u}})_i^{n+\frac{1}{2}})\Delta t,$$

$$\int_{t^n}^{t^{n+1}} f((\Delta\mathbf{u}+\tilde{\mathbf{u}})(x_{i+1},t))dt \approx f((\Delta\mathbf{u}+\tilde{\mathbf{u}})_{i+1}^{n+\frac{1}{2}})\Delta t.$$

Plugging these integrals in equation (2.14), leads to:

$$\Delta\mathbf{u}_{i+\frac{1}{2}}^{n+1} = \Delta\mathbf{u}_{i+\frac{1}{2}}^n - \frac{\Delta t}{\Delta x}[f(\Delta\mathbf{u}_{i+1}^{n+\frac{1}{2}} + \tilde{\mathbf{u}}_{i+1}) - f(\tilde{\mathbf{u}}_{i+1}) - f(\Delta\mathbf{u}_i^{n+\frac{1}{2}} + \tilde{\mathbf{u}}_i) + f(\tilde{\mathbf{u}}_i)]$$

$$+ \frac{1}{\Delta x}\int_{t^n}^{t^{n+1}} \int_{x_i}^{x_{i+1}} S(\Delta\mathbf{u},x)dxdt.$$

$$(2.15)$$

The forward projection step $(\Delta\mathbf{u}_{i+\frac{1}{2}}^n)$ of $\Delta\mathbf{u}_i^n$ onto the staggered grid is calculated using Taylor expansion of $\Delta\mathbf{u}(x,t^n)$ in space, using the fact that $\Delta\mathbf{u}(x,t^n)$ is approximated by a linear function $\mathcal{L}_i(x,t^n)$ i.e. $\Delta\mathbf{u}(x,t^n) \approx \mathcal{L}_i(x,t^n)$ in the cells of centers x_i

and x_{i+1},

$$\int_{x_i}^{x_{i+1}} \Delta\mathbf{u}(x,t^n)dx = \int_{x_i}^{x_{i+\frac{1}{2}}} \Delta\mathbf{u}(x,t^n)dx + \int_{x_{i+\frac{1}{2}}}^{x_{i+1}} \Delta\mathbf{u}(x,t^n)dx,$$

$$= \frac{\Delta x}{2}\mathcal{L}_i(x_{i+\frac{1}{4}},t^n) + \frac{\Delta x}{2}\mathcal{L}_i(x_{i+\frac{3}{4}},t^n),$$

$$= \frac{\Delta x}{2}\left(\Delta\mathbf{u}_i^n + (x_{i+\frac{1}{4}} - x_i)\frac{(\Delta\mathbf{u}_i^n)'}{\Delta x}\right)$$

$$+ \frac{\Delta x}{2}\left(\Delta\mathbf{u}_{i+1}^n + (x_{i+\frac{3}{4}} - x_{i+1})\frac{(\Delta\mathbf{u}_{i+1}^n)'}{\Delta x}\right),$$

$$= \frac{\Delta x}{2}\left(\Delta\mathbf{u}_i^n + \Delta\mathbf{u}_{i+1}^n\right) + \frac{\Delta x}{8}\left((\Delta\mathbf{u}_i^n)' - (\Delta\mathbf{u}_{i+1}^n)'\right).$$

Hence,

$$\Delta\mathbf{u}_{i+\frac{1}{2}}^n = \frac{1}{2}\left(\Delta\mathbf{u}_i^n + \Delta\mathbf{u}_{i+1}^n\right) + \frac{1}{8}\left((\Delta\mathbf{u}_i^n)' - (\Delta\mathbf{u}_{i+1}^n)'\right), \tag{2.16}$$

The predicted values $\Delta\mathbf{u}_i^{n+\frac{1}{2}}$ appearing in equation (2.15) are obtained at the intermediate time $t^{n+\frac{1}{2}}$ using a first-order Taylor expansion in time and the balance law (2.3).

The first-order Taylor expansion in time is:
$\Delta\mathbf{u}(x,t) \approx \Delta\mathbf{u}(x,a) + (t-a)\Delta\mathbf{u}_t(x,a),$ for any a and t.
For a specific point x_i,
$\Delta\mathbf{u}(x_i,t) \approx \Delta\mathbf{u}(x_i,a) + (t-a)\Delta\mathbf{u}_t(x_i,a).$
Let $a = t^n$,
$\Delta\mathbf{u}(x_i,t) \approx \Delta\mathbf{u}(x_i,t^n) + (t-t^n)\Delta\mathbf{u}_t(x_i,t^n),$
then let $t = t^{n+\frac{1}{2}}$.
Hence,

$$\Delta\mathbf{u}(x_i,t^{n+\frac{1}{2}}) \approx \Delta\mathbf{u}(x_i,t^n) + \frac{\Delta t}{2}\Delta\mathbf{u}_t(x_i,t^n),$$

$$\Delta\mathbf{u}_i^{n+\frac{1}{2}} \approx \Delta\mathbf{u}_i^n + \frac{\Delta t}{2}[-[f(\Delta\mathbf{u}+\tilde{\mathbf{u}}) - f(\tilde{\mathbf{u}})]_x|_{(x_i,t^n)} + [S(\Delta\mathbf{u},x)]|_{(x_i,t^n)}],$$

which can be written as,

$$\Delta\mathbf{u}_i^{n+\frac{1}{2}} = \Delta\mathbf{u}_i^n + \frac{\Delta t}{2}\left[-\frac{(f_i^n)'}{\Delta x} + \frac{\tilde{f}_i'}{\Delta x} + S_i^n\right] \tag{2.17}$$

where $\frac{(f_i^n)'}{\Delta x}$ and $\frac{\tilde{f}_i'}{\Delta x}$ are approximate flux derivatives with $(f_i^n)' = J_{f_i^n}.(\mathbf{u}_i^n)'$ and $\tilde{f}_i' = J_{\tilde{f}_i'}.\tilde{\mathbf{u}}_i'$. Here also, $(\mathbf{u}_i^n)'$ and $\tilde{\mathbf{u}}_i'$ are approximated by the (MC-θ) limiter (2.9). S_i^n is the discretized source term at time t^n.

On the other hand, the integral of the source term in (2.15) is discretized using the midpoint quadrature rule with respect to time and space,

$$\int_{t^n}^{t^{n+1}} \int_{x_i}^{x_{i+1}} S(\Delta\mathbf{u}, x)dxdt \approx \Delta t \int_{x_i}^{x_{i+1}} S(\Delta\mathbf{u}^{n+\frac{1}{2}}, x)dx,$$

$$\approx \Delta t \Delta x \left[\frac{S(\Delta\mathbf{u}_i^{n+\frac{1}{2}}) + S(\Delta\mathbf{u}_{i+1}^{n+\frac{1}{2}})}{2} \right].$$

Finally, the projection step $(\Delta\mathbf{u}_i^{n+1})$ of $\Delta\mathbf{u}_{i+\frac{1}{2}}^{n+1}$ back onto the original grid is calculated using Taylor expansions in space in the same way the forward projection step (2.16) was computed,

$$\Delta\mathbf{u}_i^{n+1} = \frac{1}{2}(\Delta\mathbf{u}_{i-\frac{1}{2}}^{n+1} + \Delta\mathbf{u}_{i+\frac{1}{2}}^{n+1}) + \frac{1}{8}((\Delta\mathbf{u}_{i-\frac{1}{2}}^{n+1})' - (\Delta\mathbf{u}_{i+\frac{1}{2}}^{n+1})'). \qquad (2.18)$$

Equation (2.18) gives the solution of the balance law at the next time on the original grid. The Geometry of the UC scheme and that of the NT scheme is given in figure 2.3. We see how both schemes avoid dealing with Riemann problems at the interfaces. The difference is that, the NT scheme evolves the solution on two grids, while the UC scheme evolves the solution on a single grid.

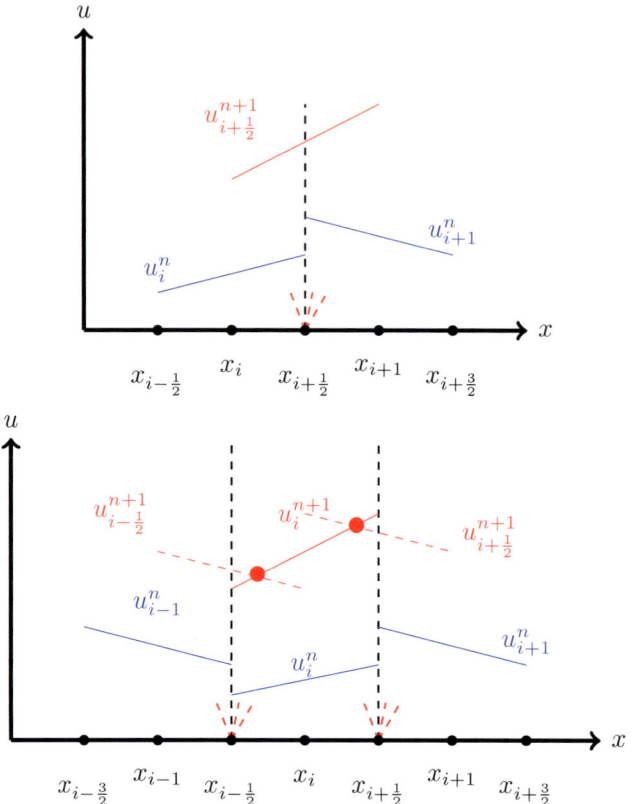

Figure 2.3: Geometry of the 1D NT scheme (left) and of the 1D UC scheme (right).

To complete the presentation of the 1D scheme, we still need to demonstrate that it is capable of capturing any stationary solution up to machine accuracy. Without any loss of generality, we assume that the updated solution satisfies $\mathbf{u}_i^n = \tilde{\mathbf{u}}_i$, i.e., $\Delta\mathbf{u}_i^n = 0$ at time $t = t^n$. Performing one iteration using the proposed numerical scheme, one can show that:

1. $\Delta\mathbf{u}_i^{n+\frac{1}{2}} = 0$.

2. $\Delta\mathbf{u}_{i+\frac{1}{2}}^{n+1} = 0$.

3. $\Delta\mathbf{u}_i^{n+1} = 0$.

The proof of 2 and 3 follows immediately after 1 is established. We start by showing 1.

The prediction step (2.17) leads to,

$$\Delta\mathbf{u}_i^{n+\frac{1}{2}} = \Delta\mathbf{u}_i^n + \frac{\Delta t}{2}\left[-\frac{f'(\Delta\mathbf{u}_i^n + \tilde{\mathbf{u}}_i)}{\Delta x} + \frac{f'(\tilde{\mathbf{u}}_i)}{\Delta x} + S(\Delta\mathbf{u}_i^n, x)\right]. \tag{2.19}$$

But since $\Delta \mathbf{u}_i^n = 0$, then we obtain,

$$\Delta \mathbf{u}_i^{n+\frac{1}{2}} = \frac{\Delta t}{2} \left[-\frac{f'(\tilde{\mathbf{u}}_i)}{\Delta x} + \frac{f'(\tilde{\mathbf{u}}_i)}{\Delta x} \right].$$

Hence, $\Delta \mathbf{u}_i^{n+\frac{1}{2}} = 0$; the proof of points 2 and 3 follows immediately. We conclude that the updated numerical solution $\mathbf{u}_i^{n+1}$ remains stationary up to machine precision.

2.3   2D Unstaggered Well-balanced FV Central Scheme

In this section we extend the proposed well-balanced scheme we derived in section 2.2 to the case of the 2D balance laws, using the subtraction technique developed in [8]. The well-balanced property of the proposed 2D scheme is presented at the end of this section. We consider the 2D balance law:

$$\begin{cases} \mathbf{U}_t + F(\mathbf{U})_x + G(\mathbf{U})_y = S(\mathbf{U}, x, y), & (x, y) \in \Omega \subset \mathbb{R}^2, \ t > 0. \\ \mathbf{U}(x, y, 0) = \mathbf{U}_0(x, y), \end{cases} \tag{2.20}$$

where $\mathbf{U}$ is the vector of conserved variables, $F(\mathbf{U})$, $G(\mathbf{U})$ are the fluxes in the x- and y- directions, respectively, and $S(\mathbf{U}, x, y)$ is the source term. We consider a Cartesian domain decomposition of the computational domain Ω where the control cells are the rectangles $C_{i,j} = \left[x_{i-\frac{1}{2}}, x_{i+\frac{1}{2}} \right] \times \left[y_{j-\frac{1}{2}}, y_{j+\frac{1}{2}} \right]$ centered at the nodes (x_i, y_j). We define the dual staggered cells $D_{i+\frac{1}{2}, j+\frac{1}{2}} = [x_i, x_{i+1}] \times [y_j, y_{j+1}]$ centered at $(x_{i+\frac{1}{2}}, y_{j+\frac{1}{2}})$. Here, $x_{i+\frac{1}{2}} = x_i + \frac{\Delta x}{2}$ and $y_{j+\frac{1}{2}} = y_j + \frac{\Delta y}{2}$, where $\Delta x = x_{i+\frac{1}{2}} - x_{i-\frac{1}{2}}$ and $\Delta y = y_{j+\frac{1}{2}} - y_{j-\frac{1}{2}}$. The visualization of the 2D grids is given in figure 2.4.

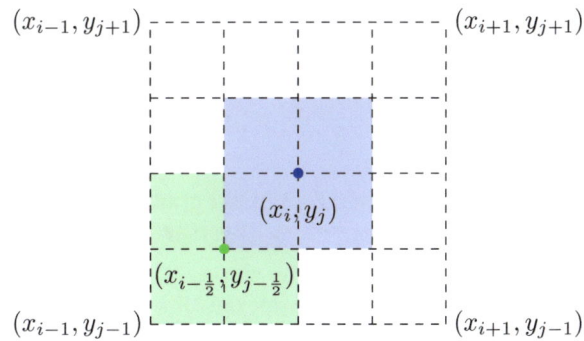

Figure 2.4: The cells of the main grid $C_{i,j}$ (blue cell) and of the staggered grid $D_{i-\frac{1}{2}, j-\frac{1}{2}}$ (green cell).

Before proceeding with the derivation of the 2D numerical method, and for convenience, we introduce the average value notations:

$$\overline{\rho}_{i,j+\frac{1}{2}} = \frac{\rho_{i,j} + \rho_{i,j+1}}{2}, \overline{\rho}_{i+\frac{1}{2},j} = \frac{\rho_{i,j} + \rho_{i+1,j}}{2}, \overline{\rho}_{i,(j)} = \frac{\rho_{i,j+\frac{1}{2}} + \rho_{i,j-\frac{1}{2}}}{2}$$

$$\bar{\rho}_{(i),j} = \frac{\rho_{i+\frac{1}{2},j} + \rho_{i-\frac{1}{2},j}}{2}, \quad [[\rho]]_{i,j+\frac{1}{2}} = \rho_{i,j+1} - \rho_{i,j}$$

$$[[\rho]]_{i+\frac{1}{2},j} = \rho_{i+1,j} - \rho_{i,j}, \ [[\rho]]_{i,(j)} = \rho_{i,j+\frac{1}{2}} - \rho_{i,j-\frac{1}{2}}, \ [[\rho]]_{(i),j} = \rho_{i+\frac{1}{2},j} - \rho_{i-\frac{1}{2},j}.$$

We follow the same strategy as in section 2.2; we assume that $\tilde{\mathbf{U}}$ is a given stationary solution of system (2.20) and we define $\Delta\mathbf{U} = \mathbf{U} - \tilde{\mathbf{U}}$. We substitute $\mathbf{U} = \Delta\mathbf{U} + \tilde{\mathbf{U}}$ in the balance law (2.20), we obtain:

$$(\Delta\mathbf{U})_t + F(\Delta\mathbf{U} + \tilde{\mathbf{U}})_x + G(\Delta\mathbf{U} + \tilde{\mathbf{U}})_y = S(\Delta\mathbf{U} + \tilde{\mathbf{U}}, x, y). \tag{2.21}$$

On the other hand, since $\tilde{\mathbf{U}}$ is a stationary solution, then balance law in (2.20) reduces to

$$F(\tilde{\mathbf{U}})_x + G(\tilde{\mathbf{U}})_y = S(\tilde{\mathbf{U}}, x, y). \tag{2.22}$$

Subtracting equation (2.22) from equation (2.21), we obtain

$$(\Delta\mathbf{U})_t + [F(\Delta\mathbf{U} + \tilde{\mathbf{U}}) - F(\tilde{\mathbf{U}})]_x + [G(\Delta\mathbf{U} + \tilde{\mathbf{U}}) - G(\tilde{\mathbf{U}})]_y = S(\Delta\mathbf{U} + \tilde{\mathbf{U}}, x, y) - S(\tilde{\mathbf{U}}, x, y). \tag{2.23}$$

Using the fact that the source term $S(\mathbf{U}, x, y)$ in equation (2.20) is linear in terms of the conserved variables, then equation (2.23) reduces to

$$(\Delta\mathbf{U})_t + [F(\Delta\mathbf{U} + \tilde{\mathbf{U}}) - F(\tilde{\mathbf{U}})]_x + [G(\Delta\mathbf{U} + \tilde{\mathbf{U}}) - G(\tilde{\mathbf{U}})]_y = S(\Delta\mathbf{U}, x, y). \tag{2.24}$$

The proposed numerical scheme consists of evolving the balance law (2.24) instead of evolving the balance law in system (2.20). The numerical solution $\mathbf{U}$ will be then obtained using the formula $\mathbf{U} = \Delta\mathbf{U} + \tilde{\mathbf{U}}$. The numerical scheme that we shall use to evolve $\Delta\mathbf{U}(x, y, t)$ follows a classical finite volume approach; it evolves a piecewise linear function $\mathcal{L}_{i,j}(x, y, t)$ defined on the control cells $C_{i,j}$ and used to approximate the analytic solution $\Delta\mathbf{U}(x, y, t)$ of system (2.20). Without any loss of generality we can assume that $\Delta\mathbf{U}_{i,j}^n$ is known at time t^n and we define $\mathcal{L}_{i,j}(x, y, t^n)$ on the cells $C_{i,j}$ as follows.

$$\mathcal{L}_{i,j}(x, y, t^n) = \Delta\mathbf{U}_{i,j}^n + (x - x_i)\frac{(\Delta\mathbf{U}_{i,j}^{n,x})'}{\Delta x} + (y - y_j)\frac{(\Delta\mathbf{U}_{i,j}^{n,y})'}{\Delta y}, \quad \forall (x, y) \in C_{i,j},$$

where $\frac{(\Delta\mathbf{U}_{i,j}^{n,x})'}{\Delta x}$ and $\frac{(\Delta\mathbf{U}_{i,j}^{n,y})'}{\Delta y}$ are limited numerical gradients approximating $\frac{\partial\Delta\mathbf{U}}{\partial x}(x, y_j, t^n)|_{x=x_i}$ and $\frac{\partial\Delta\mathbf{U}}{\partial y}(x_i, y, t^n)|_{y=y_j}$, respectively, at the point (x_i, y_j, t^n). The (MC-θ) limiter (2.9) is used to compute the slopes $(\Delta\mathbf{U}_{i,j}^{n,x})'$ and $(\Delta\mathbf{U}_{i,j}^{n,y})'$ in order to avoid spurious oscillations. Next, we integrate the balance law (2.24) over the rectangular box $R_{i+\frac{1}{2},j+\frac{1}{2}}^n = D_{i+\frac{1}{2},j+\frac{1}{2}} \times [t^n, t^{n+1}]$,

$$\iiint_{R_{i+\frac{1}{2},j+\frac{1}{2}}^n} (\Delta\mathbf{U})_t + [F(\Delta\mathbf{U} + \tilde{\mathbf{U}}) - F(\tilde{\mathbf{U}})]_x + [G(\Delta\mathbf{U} + \tilde{\mathbf{U}}) - G(\tilde{\mathbf{U}})]_y dR$$

$$= \iiint_{R_{i+\frac{1}{2},j+\frac{1}{2}}^n} S(\Delta\mathbf{U}, x, y) dR. \tag{2.25}$$

We use the fact that $\Delta\mathbf{U}$ is approximated using piecewise linear interpolants similar to $\mathcal{L}_{i,j}$ on the cells $C_{i,j}$; following the derivation of the unstaggered central schemes in [73], equation (2.25) is rewritten as:

$$
\Delta\mathbf{U}^{n+1}_{i+\frac{1}{2},j+\frac{1}{2}} = \Delta\mathbf{U}^{n}_{i+\frac{1}{2},j+\frac{1}{2}} - \frac{1}{\Delta x \Delta y}\iiint_{R^n_{i+\frac{1}{2},j+\frac{1}{2}}} [F(\Delta\mathbf{U}+\tilde{\mathbf{U}}) - F(\tilde{\mathbf{U}})]_x
$$
$$
+ [G(\Delta\mathbf{U}+\tilde{\mathbf{U}}) - G(\tilde{\mathbf{U}})]_y\, dR + \frac{1}{\Delta x \Delta y}\iiint_{R^n_{i+\frac{1}{2},j+\frac{1}{2}}} S(\Delta\mathbf{U},x,y)dR. \quad (2.26)
$$

For the flux integrals, we apply the divergence theorem that changes the volume integral into surface integral. Equation (2.26) becomes then:

$$
\Delta\mathbf{U}^{n+1}_{i+\frac{1}{2},j+\frac{1}{2}} = \Delta\mathbf{U}^{n}_{i+\frac{1}{2},j+\frac{1}{2}} - \frac{1}{\Delta x \Delta y}\int_{t^n}^{t^{n+1}}\int_{\partial D_{i+\frac{1}{2},j+\frac{1}{2}}} [F(\Delta\mathbf{U}+\tilde{\mathbf{U}}) - F(\tilde{\mathbf{U}})]\cdot n_x dA dt
$$
$$
- \frac{1}{\Delta x \Delta y}\int_{t^n}^{t^{n+1}}\int_{\partial D_{i+\frac{1}{2},j+\frac{1}{2}}} [G(\Delta\mathbf{U}+\tilde{\mathbf{U}}) - G(\tilde{\mathbf{U}})]\cdot n_y dA dt
$$
$$
+ \frac{1}{\Delta x \Delta y}\iiint_{R^n_{i+\frac{1}{2},j+\frac{1}{2}}} S(\Delta\mathbf{U},x,y)dR \quad (2.27)
$$

where $\mathbf{n} = (n_x, n_y)$ is the outward pointing unit normal at each point on the boundary $\partial D_{i+\frac{1}{2},j+\frac{1}{2}}$ (the boundary of $D_{i+\frac{1}{2},j+\frac{1}{2}}$), see figure 2.5.

Next, we approximate the integrals

$$
I = \int_{t^n}^{t^{n+1}}\int_{\partial D_{i+\frac{1}{2},j+\frac{1}{2}}} [F(\Delta\mathbf{U}+\tilde{\mathbf{U}}) - F(\tilde{\mathbf{U}})].n_x dx dy dt
$$

and

$$
J = \int_{t^n}^{t^{n+1}}\int_{\partial D_{i+\frac{1}{2},j+\frac{1}{2}}} [G(\Delta\mathbf{U}+\tilde{\mathbf{U}}) - G(\tilde{\mathbf{U}})].n_y dx dy dt.
$$

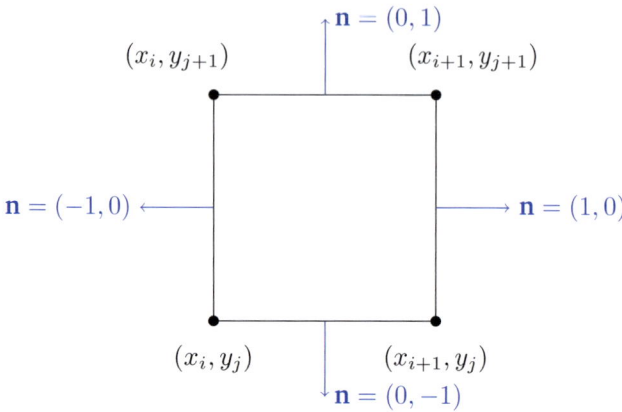

Figure 2.5: The boundary $\partial D_{i+\frac{1}{2},j+\frac{1}{2}}$ and the outward pointing unit normal vector $\mathbf{n} = (n_x, n_y)$ on each side of the boundary.

$$I = \int_{t^n}^{t^{n+1}} \int_{\partial D_{i+\frac{1}{2},j+\frac{1}{2}}} [F(\Delta\mathbf{U} + \tilde{\mathbf{U}}) - F(\tilde{\mathbf{U}})] \cdot n_x dx dy dt$$

$$= \int_{t^n}^{t^{n+1}} \int_{y_j}^{y_{j+1}} [F((\Delta\mathbf{U} + \tilde{\mathbf{U}})(x_{i+1}, y, t)) - F(\tilde{\mathbf{U}}(x_{i+1}, y, t))] \cdot 1 dy$$

$$+ \int_{t^n}^{t^{n+1}} \int_{x_{i+1}}^{x_i} [F((\Delta\mathbf{U} + \tilde{\mathbf{U}})(x, y_{j+1}, t)) - F(\tilde{\mathbf{U}}(x, y_{j+1}, t))] \cdot 0 dx$$

$$+ \int_{t^n}^{t^{n+1}} \int_{y_{j+1}}^{y_j} [F((\Delta\mathbf{U} + \tilde{\mathbf{U}})(x_i, y, t)) - F(\tilde{\mathbf{U}}(x_i, y, t))] \cdot -1 dy$$

$$+ \int_{t^n}^{t^{n+1}} \int_{x_i}^{x_{i+1}} [F((\Delta\mathbf{U} + \tilde{\mathbf{U}})(x, y_j, t)) - F(\tilde{\mathbf{U}}(x, y_j, t))] \cdot 0 dx$$

An approximation of each integral using the midpoint rule leads to:

$$I = \frac{\Delta t \Delta y}{2} \Big[[F((\Delta\mathbf{U} + \tilde{\mathbf{U}})(x_{i+1}, y_j, t^{n+\frac{1}{2}})) - F(\tilde{\mathbf{U}}(x_{i+1}, y_j, t^{n+\frac{1}{2}}))]$$

$$+ [F((\Delta\mathbf{U} + \tilde{\mathbf{U}})(x_{i+1}, y_{j+1}, t^{n+\frac{1}{2}})) - F(\tilde{\mathbf{U}}(x_{i+1}, y_{j+1}, t^{n+\frac{1}{2}}))] \Big]$$

$$\frac{\Delta t \Delta y}{2} \Big[[F((\Delta\mathbf{U} + \tilde{\mathbf{U}})(x_i, y_j, t^{n+\frac{1}{2}})) - F(\tilde{\mathbf{U}}(x_i, y_j, t^{n+\frac{1}{2}}))]$$

$$+ [F((\Delta\mathbf{U} + \tilde{\mathbf{U}})(x_i, y_{j+1}, t^{n+\frac{1}{2}})) - F(\tilde{\mathbf{U}}(x_i, y_{j+1}, t^{n+\frac{1}{2}}))] \Big].$$

Hence,

$$
I = \frac{\Delta t \Delta y}{2} \Big[[F((\Delta \mathbf{U} + \tilde{\mathbf{U}})_{i+1,j}^{n+\frac{1}{2}}) - F(\tilde{\mathbf{U}}_{i+1,j}^{n+\frac{1}{2}})] + [F((\Delta \mathbf{U} + \tilde{\mathbf{U}})_{i+1,j+1}^{n+\frac{1}{2}}) - F(\tilde{\mathbf{U}}_{i+1,j+1}^{n+\frac{1}{2}})]
$$
$$
- [F((\Delta \mathbf{U} + \tilde{\mathbf{U}})_{i,j}^{n+\frac{1}{2}}) - F(\tilde{\mathbf{U}}_{i,j}^{n+\frac{1}{2}})] - [F((\Delta \mathbf{U} + \tilde{\mathbf{U}})_{i,j+1}^{n+\frac{1}{2}}) - F(\tilde{\mathbf{U}}_{i,j+1}^{n+\frac{1}{2}})] \Big].
$$

Similar approximation for J implies,

$$
J = \frac{\Delta t \Delta x}{2} \Big[[G((\Delta \mathbf{U} + \tilde{\mathbf{U}})_{i,j+1}^{n+\frac{1}{2}}) - G(\tilde{\mathbf{U}}_{i,j+1}^{n+\frac{1}{2}})] + [G((\Delta \mathbf{U} + \tilde{\mathbf{U}})_{i+1,j+1}^{n+\frac{1}{2}}) - G(\tilde{\mathbf{U}}_{i+1,j+1}^{n+\frac{1}{2}})]
$$
$$
- [G((\Delta \mathbf{U} + \tilde{\mathbf{U}})_{i,j}^{n+\frac{1}{2}}) - G(\tilde{\mathbf{U}}_{i,j}^{n+\frac{1}{2}})] - [G((\Delta \mathbf{U} + \tilde{\mathbf{U}})_{i+1,j}^{n+\frac{1}{2}}) - G(\tilde{\mathbf{U}}_{i+1,j}^{n+\frac{1}{2}})] \Big].
$$

Hence equation (2.27) becomes,

$$
\Delta \mathbf{U}_{i+\frac{1}{2},j+\frac{1}{2}}^{n+1} = \Delta \mathbf{U}_{i+\frac{1}{2},j+\frac{1}{2}}^{n}
$$
$$
- \frac{\Delta t}{2\Delta x} \Big[[F((\Delta \mathbf{U} + \tilde{\mathbf{U}})_{i+1,j}^{n+\frac{1}{2}}) - F(\tilde{\mathbf{U}}_{i+1,j}^{n+\frac{1}{2}})] + [F((\Delta \mathbf{U} + \tilde{\mathbf{U}})_{i+1,j+1}^{n+\frac{1}{2}}) - F(\tilde{\mathbf{U}}_{i+1,j+1}^{n+\frac{1}{2}})]
$$
$$
- [F((\Delta \mathbf{U} + \tilde{\mathbf{U}})_{i,j}^{n+\frac{1}{2}}) - F(\tilde{\mathbf{U}}_{i,j}^{n+\frac{1}{2}})] - [F((\Delta \mathbf{U} + \tilde{\mathbf{U}})_{i,j+1}^{n+\frac{1}{2}}) - F(\tilde{\mathbf{U}}_{i,j+1}^{n+\frac{1}{2}})] \Big]
$$
$$
- \frac{\Delta t}{2\Delta y} \Big[[G((\Delta \mathbf{U} + \tilde{\mathbf{U}})_{i,j+1}^{n+\frac{1}{2}}) - G(\tilde{\mathbf{U}}_{i,j+1}^{n+\frac{1}{2}})] + [G((\Delta \mathbf{U} + \tilde{\mathbf{U}})_{i+1,j+1}^{n+\frac{1}{2}}) - G(\tilde{\mathbf{U}}_{i+1,j+1}^{n+\frac{1}{2}})]
$$
$$
- [G((\Delta \mathbf{U} + \tilde{\mathbf{U}})_{i,j}^{n+\frac{1}{2}}) - G(\tilde{\mathbf{U}}_{i,j}^{n+\frac{1}{2}})] - [G((\Delta \mathbf{U} + \tilde{\mathbf{U}})_{i+1,j}^{n+\frac{1}{2}}) - G(\tilde{\mathbf{U}}_{i+1,j}^{n+\frac{1}{2}})] \Big]
$$
$$
+ \frac{1}{\Delta x \Delta y} \iiint_{R_{i+\frac{1}{2},j+\frac{1}{2}}^{n}} S(\Delta \mathbf{U}, x, y) dR. \quad (2.28)
$$

The integral of the source term is being approximated using the midpoint quadrature rule both in time and space:

$$
\iiint_{R_{i+\frac{1}{2},j+\frac{1}{2}}^{n}} S(\Delta \mathbf{U}, x, y) dR \approx \Delta t \Delta x \Delta y
$$
$$
\left[\frac{S(\Delta \mathbf{U}_{i,j}^{n+\frac{1}{2}}) + S(\Delta \mathbf{U}_{i+1,j}^{n+\frac{1}{2}}) + S(\Delta \mathbf{U}_{i,j+1}^{n+\frac{1}{2}}) + S(\Delta \mathbf{U}_{i+1,j+1}^{n+\frac{1}{2}})}{4} \right]. \quad (2.29)
$$

The forward projection step in equation (2.28) consists of projecting the solution at time t^n onto the staggered grid. It is performed using linear interpolations in two space dimensions in addition to Taylor expansions in space; we obtain:

$$
\Delta \mathbf{U}_{i+\frac{1}{2},j+\frac{1}{2}}^{n} = \frac{1}{2}(\overline{\Delta \mathbf{U}}_{i+\frac{1}{2},j}^{n} + \overline{\Delta \mathbf{U}}_{i+\frac{1}{2},j+1}^{n})
$$
$$
- \frac{1}{16}([[\Delta \mathbf{U}^{n,x}]]_{i+\frac{1}{2},j} + [[\Delta \mathbf{U}^{n,x}]]_{i+\frac{1}{2},j+1})
$$
$$
- \frac{1}{16}([[\Delta \mathbf{U}^{n,y}]]_{i,j+\frac{1}{2}} + [[\Delta \mathbf{U}^{n,y}]]_{i+1,j+\frac{1}{2}}). \quad (2.30)
$$

Here, $\Delta \mathbf{U}^{n,x}$ and $\Delta \mathbf{U}^{n,y}$ are the spatial partial derivatives of $\Delta \mathbf{U}^n$ that are approximated using the (MC-θ) limiter (2.9).

Finally, the evolution step (2.28) at time t^{n+1} on the staggered nodes can be written as,

$$
\begin{aligned}
\Delta \mathbf{U}^{n+1}_{i+\frac{1}{2},j+\frac{1}{2}} = \; & \Delta \mathbf{U}^n_{i+\frac{1}{2},j+\frac{1}{2}} \\
& - \frac{\Delta t}{2}[D^x_+ F(\Delta \mathbf{U}^{n+\frac{1}{2}}_{i,j} + \tilde{\mathbf{U}}_{i,j}) - D^x_+ F(\tilde{\mathbf{U}}_{i,j}) + D^x_+ F(\Delta \mathbf{U}^{n+\frac{1}{2}}_{i,j+1} + \tilde{\mathbf{U}}_{i,j+1}) \\
& \qquad\qquad - D^x_+ F(\tilde{\mathbf{U}}_{i,j+1})] \\
& - \frac{\Delta t}{2}[D^y_+ G(\Delta \mathbf{U}^{n+\frac{1}{2}}_{i,j} + \tilde{\mathbf{U}}_{i,j}) - D^y_+ G(\tilde{\mathbf{U}}_{i,j}) + D^y_+ F(\Delta \mathbf{U}^{n+\frac{1}{2}}_{i+1,j} + \tilde{\mathbf{U}}_{i+1,j}) \\
& \qquad\qquad - D^y_+ G(\tilde{\mathbf{U}}_{i+1,j})] \\
& + \Delta t . S(\Delta \mathbf{U}^{n+\frac{1}{2}}_{i,j}, \Delta \mathbf{U}^{n+\frac{1}{2}}_{i+1,j}, \Delta \mathbf{U}^{n+\frac{1}{2}}_{i,j+1}, \Delta \mathbf{U}^{n+\frac{1}{2}}_{i+1,j+1}). \quad (2.31)
\end{aligned}
$$

With

$$
\begin{aligned}
S(\Delta \mathbf{U}^{n+\frac{1}{2}}_{i,j}, &\Delta \mathbf{U}^{n+\frac{1}{2}}_{i+1,j}, \Delta \mathbf{U}^{n+\frac{1}{2}}_{i,j+1}, \Delta \mathbf{U}^{n+\frac{1}{2}}_{i+1,j+1}) = \\
& \left[\frac{S(\Delta \mathbf{U}^{n+\frac{1}{2}}_{i,j}) + S(\Delta \mathbf{U}^{n+\frac{1}{2}}_{i+1,j}) + S(\Delta \mathbf{U}^{n+\frac{1}{2}}_{i,j+1}) + S(\Delta \mathbf{U}^{n+\frac{1}{2}}_{i+1,j+1})}{4} \right].
\end{aligned}
$$

Here D^x_+ and D^y_+ are the forward differences given by,
$$D^x_+ F(\mathbf{U}_{i,j}) = \frac{F(\mathbf{U}_{i+1,j}) - F(\mathbf{U}_{i,j})}{\Delta x}, D^y_+ F(\mathbf{U}_{i,j}) = \frac{F(\mathbf{U}_{i,j+1}) - F(\mathbf{U}_{i,j})}{\Delta y}.$$

The predicted values in equation (2.31) are generated at time $t^{n+\frac{1}{2}}$ using a first order Taylor expansion in time in addition to the balance law (2.20):

$$
\Delta \mathbf{U}^{n+\frac{1}{2}}_{i,j} = \Delta \mathbf{U}^n_{i,j} + \frac{\Delta t}{2} \left[-\frac{(F^n_{i,j})'}{\Delta x} + \frac{\tilde{F}'_{i,j}}{\Delta x} - \frac{(G^n_{i,j})'}{\Delta y} + \frac{\tilde{G}'_{i,j}}{\Delta y} + S^n_{i,j} \right], \qquad (2.32)
$$

where $\frac{(F^n_{i,j})'}{\Delta x}, \frac{\tilde{F}'_{i,j}}{\Delta x}, \frac{(G^n_{i,j})'}{\Delta y}$ and $\frac{\tilde{G}'_{i,j}}{\Delta y}$ denote the approximate flux derivatives with $(F^n_{i,j})' = J_{F^n_{i,j}}.\mathbf{U}^{n,x}_{i,j}, \tilde{F}'_{i,j} = J_{\tilde{F}_{i,j}}.\tilde{\mathbf{U}}^x_{i,j}, (G^n_{i,j})' = J_{G^n_{i,j}}.\mathbf{U}^{n,y}_{i,j}, \tilde{G}'_{i,j} = J_{\tilde{G}_{i,j}}.\tilde{\mathbf{U}}^y_{i,j}$. Here, we also use the (MC-θ) limiter (2.9) to compute the slopes $\mathbf{U}^{n,x}_{i,j}, \tilde{\mathbf{U}}^x_{i,j}, \mathbf{U}^{n,y}_{i,j}$, and $\tilde{\mathbf{U}}^y_{i,j}$ in order to avoid spurious oscillations. $S^n_{i,j}$ is the discrete source term.

Finally we apply a back projection step similar to the one in (2.30). In order to retrieve the solution at the time t^{n+1} on the original cells $C_{i,j}$, we obtain

$$
\begin{aligned}
\Delta \mathbf{U}^{n+1}_{i,j} = \; & \frac{1}{2}(\overline{\Delta \mathbf{U}}^{n+1}_{i,j-\frac{1}{2}} + \overline{\Delta \mathbf{U}}^{n+1}_{i,j+\frac{1}{2}}) \\
& - \frac{1}{16}([[\Delta \mathbf{U}^{n+1,x}]]_{(i),j-\frac{1}{2}} + [[\Delta \mathbf{U}^{n+1,x}]]_{(i),j+\frac{1}{2}}) \\
& - \frac{1}{16}([[\Delta \mathbf{U}^{n+1,y}]]_{i-\frac{1}{2},(j)} + [[\Delta \mathbf{U}^{n+1,y}]]_{i+\frac{1}{2},(j)}), \quad (2.33)
\end{aligned}
$$

where $\Delta \mathbf{U}_{i,j}^{n+1,x}$ and $\Delta \mathbf{U}_{i,j}^{n+1,y}$ denote the spatial partial derivatives of the numerical solution obtained at time t^{n+1} and node (x_i, y_j) approximated using the (MC-θ) limiter (2.9).

To complete the presentation of the numerical scheme, we need to verify the well-balanced property of the proposed scheme and to show that it is capable of maintaining stationary solutions of the Euler system with gravitational source term.

Suppose that the numerical solution obtained at time $t = t^n$ satisfies $\mathbf{U}_{i,j}^n = \tilde{\mathbf{U}}_{i,j}$, i.e., $\Delta \mathbf{U}_{i,j}^n = 0$. Performing one iteration using the proposed numerical scheme, one can show that:

1. $\Delta \mathbf{U}_{i,j}^{n+\frac{1}{2}} = 0$.

2. $\Delta \mathbf{U}_{i+\frac{1}{2},j+\frac{1}{2}}^{n+1} = 0$.

3. $\Delta \mathbf{U}_{i,j}^{n+1} = 0$.

In fact, it is straight forward to establish 2 and 3 once 1 is established. We will present the proof of 1 only.

The prediction step (2.32) leads to

$$\Delta \mathbf{U}_{i,j}^{n+\frac{1}{2}} = \Delta \mathbf{U}_{i,j}^n + \frac{\Delta t}{2} \left[-\frac{F'(\Delta \mathbf{U}_{i,j}^n + \tilde{\mathbf{U}}_{i,j})}{\Delta x} + \frac{F'(\tilde{\mathbf{U}}_{i,j})}{\Delta x} \right.$$
$$\left. -\frac{G'(\Delta \mathbf{U}_{i,j}^n + \tilde{\mathbf{U}}_{i,j})}{\Delta y} + \frac{G'(\tilde{\mathbf{U}}_{i,j})}{\Delta y} + S(\Delta \mathbf{U}_{i,j}^n, x, y) \right]. \quad (2.34)$$

But since $\Delta \mathbf{U}_{i,j}^n = 0$, then we obtain,

$$\Delta \mathbf{U}_{i,j}^{n+\frac{1}{2}} = \frac{\Delta t}{2} \left[-\frac{F'(\tilde{\mathbf{U}}_{i,j})}{\Delta x} + \frac{F'(\tilde{\mathbf{U}}_{i,j})}{\Delta x} - \frac{G'(\tilde{\mathbf{U}}_{i,j})}{\Delta y} + \frac{G'(\tilde{\mathbf{U}}_{i,j})}{\Delta y} \right].$$

Hence, $\Delta \mathbf{U}_{i,j}^{n+\frac{1}{2}} = 0$. Therefore, we conclude that the updated numerical solution remains stationary up to machine precision.

2.4 TVD Property of the Scheme Applied to Scalar Conservation Law

In this section, we establish the Total Variation Diminishing (TVD) property of our proposed numerical schemes. To prove that the scheme is TVD, one needs to prove that $TV(u(t + \Delta t)) \leq TV(u(t))$.

Let the scalar conservation law,

$$u_t + f(u)_x = 0. \quad (2.35)$$

with $f(u)_x = a(u)u_x$. As in (2.1) and (2.20), we will discretize the equation,

$$\Delta u_t + h(\Delta u)_x = 0, \tag{2.36}$$

where $\Delta u = u - \tilde{u}$ and $h(\Delta u) = f(\Delta u + \tilde{u}) - f(\tilde{u})$ and $\tilde{u}$ a time independent reference solution. Using our unstaggered central scheme, the numerical solution of the scalar equation (2.36) is updated at time t^{n+1} as follows: First, we apply a forward projection step,

$$\Delta u_{i+\frac{1}{2}}^n = \frac{1}{2}\left(\Delta u_i^n + \Delta u_{i+1}^n\right) + \frac{1}{8}\left(\left(\Delta u_i^n\right)' - \left(\Delta u_{i+1}^n\right)'\right). \tag{2.37}$$

Then, we predict the solution values at time $t^{n+\frac{1}{2}}$ with the aid of the predictor step,

$$\Delta u_i^{n+\frac{1}{2}} = \Delta u_i^n - \frac{\Delta t}{2}\left[\frac{(h_i^n)'}{\Delta x}\right]. \tag{2.38}$$

Next, we apply the time evolution step

$$\Delta u_{i+\frac{1}{2}}^{n+1} = \Delta u_{i+\frac{1}{2}}^n - \lambda\left[h(\Delta u_{i+1}^{n+\frac{1}{2}}) - h(\Delta u_i^{n+\frac{1}{2}})\right], \tag{2.39}$$

with $\lambda = \frac{\Delta t}{\Delta x}$. Finally, we apply the backward projection step

$$\Delta u_i^{n+1} = \frac{1}{2}\left(\Delta u_{i-\frac{1}{2}}^{n+1} + \Delta u_{i+\frac{1}{2}}^{n+1}\right) + \frac{1}{8}\left(\left(\Delta u_{i-\frac{1}{2}}^{n+1}\right)' - \left(\Delta u_{i+\frac{1}{2}}^{n+1}\right)'\right). \tag{2.40}$$

Theorem 1. *Assume that the numerical spatial derivatives be chosen as,*

$$0 \leq \Delta u_i'.sgn(\Delta u_{i+1} - \Delta u_i) \leq Cst_{\Delta u}.|minmod\left(\Delta u_{i+1} - \Delta u_i, \Delta u_i - \Delta u_{i-1}\right)|,$$

$$0 \leq h_i'.sgn(\Delta u_{i+1} - \Delta u_i) \leq Cst_h.|minmod\left(\Delta u_{i+1} - \Delta u_i, \Delta u_i - \Delta u_{i-1}\right)|,$$

with $Cst_{\Delta u} = \alpha$ and the following CFL condition holds,

$$\lambda.max|a(u_i)| \leq \beta,$$

where

$$\beta = \lambda\frac{Cst_h}{Cst_{\Delta u}} \leq \frac{\sqrt{4 + 4\alpha - \alpha^2} - 2}{2\alpha},$$

and $\alpha < 4$ (for $\beta > 0$). Then the scheme satisfies the TVD property.

Proof. Inspired by the TVD proof in [61] and [39], one can say that it is sufficient to prove that $|A_i| \leq \frac{1}{2}$ and $|C_{i+\frac{1}{2}}| \leq \frac{1}{2}$ with $A_i = \dfrac{\frac{1}{8}\left(\left(\Delta u_{i-\frac{1}{2}}^{n+1}\right)' - \left(\Delta u_{i+\frac{1}{2}}^{n+1}\right)'\right)}{\left(\Delta u_{i+\frac{1}{2}}^{n+1} - \Delta u_{i-\frac{1}{2}}^{n+1}\right)}$ and

$$C_{i+\frac{1}{2}} = \frac{\lambda \left[h(\Delta u_{i+1}^{n+\frac{1}{2}}) - h(\Delta u_i^{n+\frac{1}{2}}) \right] - \frac{1}{8} \left((\Delta u_i^n)' - (\Delta u_{i+1}^n)' \right)}{\Delta u_{i+1}^n - \Delta u_i^n}.$$

First, we show that $|A_i| \leq \frac{1}{2}$,

$$\frac{1}{8} \left| \frac{\left(\Delta u_{i-\frac{1}{2}}^{n+1} \right)' - \left(\Delta u_{i+\frac{1}{2}}^{n+1} \right)'}{\left(\Delta u_{i+\frac{1}{2}}^{n+1} - \Delta u_{i-\frac{1}{2}}^{n+1} \right)} \right|$$

$$\leq \frac{1}{8} \max \left(\left| \frac{\left(\Delta u_{i-\frac{1}{2}}^{n+1} \right)'}{\left(\Delta u_{i+\frac{1}{2}}^{n+1} - \Delta u_{i-\frac{1}{2}}^{n+1} \right)} \right|, \left| \frac{\left(\Delta u_{i+\frac{1}{2}}^{n+1} \right)'}{\left(\Delta u_{i+\frac{1}{2}}^{n+1} - \Delta u_{i-\frac{1}{2}}^{n+1} \right)} \right| \right) \leq \frac{\alpha}{8} \leq \frac{1}{2}. \quad (2.41)$$

Next, we show that $|C_{i+\frac{1}{2}}| \leq \frac{1}{2}$,

$$\left| \frac{\lambda \left[h(\Delta u_{i+1}^{n+\frac{1}{2}}) - h(\Delta u_i^{n+\frac{1}{2}}) \right] - \frac{1}{8} \left((\Delta u_i^n)' - (\Delta u_{i+1}^n)' \right)}{\Delta u_{i+1}^n - \Delta u_i^n} \right|$$

$$\leq \lambda \left| \frac{h(\Delta u_{i+1}^{n+\frac{1}{2}}) - h(\Delta u_i^{n+\frac{1}{2}})}{\Delta u_{i+1}^n - \Delta u_i^n} \right| + \frac{1}{8} \left| \frac{(\Delta u_i^n)' - (\Delta u_{i+1}^n)'}{\Delta u_{i+1}^n - \Delta u_i^n} \right|$$

$$\leq \lambda \left| \frac{h(\Delta u_{i+1}^{n+\frac{1}{2}}) - h(\Delta u_i^{n+\frac{1}{2}})}{\Delta u_{i+1}^{n+\frac{1}{2}} - \Delta u_i^{n+\frac{1}{2}}} \right| \cdot \left| \frac{\Delta u_{i+1}^{n+\frac{1}{2}} - \Delta u_i^{n+\frac{1}{2}}}{\Delta u_{i+1}^n - \Delta u_i^n} \right| + \frac{1}{8} \left| \frac{(\Delta u_i^n)' - (\Delta u_{i+1}^n)'}{\Delta u_{i+1}^n - \Delta u_i^n} \right|.$$

$$(2.42)$$

From the CFL condition, one concludes that,

$$\lambda \left| \frac{h(\Delta u_{i+1}^{n+\frac{1}{2}}) - h(\Delta u_i^{n+\frac{1}{2}})}{\Delta u_{i+1}^{n+\frac{1}{2}} - \Delta u_i^{n+\frac{1}{2}}} \right| \leq \beta. \quad (2.43)$$

Next, from the predictor step $\Delta u_i^{n+\frac{1}{2}}$, the second absolute value to the right-hand side of inequality (2.42) is bounded by

$$\left| \frac{\Delta u_{i+1}^{n+\frac{1}{2}} - \Delta u_i^{n+\frac{1}{2}}}{\Delta u_{i+1}^n - \Delta u_i^n} \right| = \left| \frac{\Delta u_{i+1}^n - \frac{\lambda}{2} h'_{i+1} - \Delta u_i^n + \frac{\lambda}{2} h'_i}{\Delta u_{i+1}^n - \Delta u_i^n} \right|$$

$$= \left| \frac{\Delta u_{i+1}^n - \Delta u_i^n - \frac{\lambda}{2} (h'_{i+1} - h'_i)}{\Delta u_{i+1}^n - \Delta u_i^n} \right|$$

$$\leq 1 + \frac{\lambda}{2} \left| \frac{h'_{i+1} - h'_i}{\Delta u_{i+1}^n - \Delta u_i^n} \right| \leq 1 + \frac{\lambda}{2} \max \left(\left| \frac{h'_{i+1}}{\Delta u_{i+1}^n - \Delta u_i^n} \right|, \left| \frac{h'_i}{\Delta u_{i+1}^n - \Delta u_i^n} \right| \right)$$

$$\leq 1 + \frac{\lambda}{2} Cst_h \leq 1 + \frac{\alpha \beta}{2}. \quad (2.44)$$

Finally, we have

$$
\frac{1}{8}\left|\frac{(\Delta u_i^n)' - (\Delta u_{i+1}^n)'}{\Delta u_{i+1}^n - \Delta u_i^n}\right| \leq \frac{1}{8}\max\left(\left|\frac{(\Delta u_{i+1}^n)'}{\Delta u_{i+1}^n - \Delta u_i^n}\right|, \left|\frac{(\Delta u_i^n)'}{\Delta u_{i+1}^n - \Delta u_i^n}\right|\right) \leq \frac{\alpha}{8}.
$$

$$(2.45)$$

Performing the following term-by-term operations, $(2.43) \times (2.44) + (2.45)$ results in,

$$
\left|\frac{\lambda\left[h(\Delta u_{i+1}^{n+\frac{1}{2}}) - h(\Delta u_i^{n+\frac{1}{2}})\right] - \frac{1}{8}\left((\Delta u_i^n)' - (\Delta u_{i+1}^n)'\right)}{\Delta u_{i+1}^n - \Delta u_i^n}\right|
$$

$$\leq \beta(1 + \frac{1}{2}\alpha\beta) + \frac{1}{8}\alpha \leq \frac{1}{2}. \quad (2.46)$$

This follows from the definition of β, and we conclude that,

$$|C_{i+\frac{1}{2}}| \leq \frac{1}{2}. \quad (2.47)$$

The total variation in the updated solution is now,

$$
TV(\Delta u(t + \Delta t)) = \sum_i |\Delta u_{i+1}(t + \Delta t) - \Delta u_i(t + \Delta t)|,
$$

$$
\leq \sum_i \left|\Delta u_{i+\frac{3}{2}}^{n+1} - \Delta u_{i+\frac{1}{2}}^{n+1}\right|\left|\frac{1}{2} + A_{i+1}\right| + \left|\Delta u_{i+\frac{1}{2}}^{n+1} - \Delta u_{i-\frac{1}{2}}^{n+1}\right|\left|\frac{1}{2} - A_i\right|,
$$

$$
= \sum_i \left|\Delta u_{i+\frac{1}{2}}^{n+1} - \Delta u_{i-\frac{1}{2}}^{n+1}\right|,
$$

$$
\leq \sum_i \left|\Delta u_{i+1}^n - \Delta u_i^n\right|\left|\frac{1}{2} - C_{i+\frac{1}{2}}\right| + \left|\Delta u_{i+1}^n - \Delta u_i^n\right|\left|\frac{1}{2} + C_{i-\frac{1}{2}}\right|,
$$

$$
= \sum_i \left|\Delta u_{i+1}^n - \Delta u_i^n\right| = \sum_i |\Delta u_{i+1}(t) - \Delta u_i(t)| = TV(\Delta u(t)),
$$

here we followed a re-indexing step twice. We conclude that

$$
TV(u(t + \Delta t)) - TV(u(t)) = TV(\Delta u(t + \Delta t) + \tilde{u}) - TV(\Delta u(t) + \tilde{u}),
$$

$$
\leq TV(\Delta u(t + \Delta t)) + TV(\tilde{u}) - TV(\Delta u(t)) - TV(\tilde{u}),
$$

$$
= TV(\Delta u(t + \Delta t)) - TV(\Delta u(t)) \leq 0.
$$

Hence,

$$
TV(u(t + \Delta t)) \leq TV(u(t)).
$$

$\square$

Theorem 1 states that the scheme is TVD in the scalar case, which assures, according to the Lax-Wendroff theorem [58], the convergence of the scheme to a weak solution of the conservation law in the scalar case.

2.5   Numerical Results

In this section, we implement the proposed well-balanced numerical schemes and use them to solve classical problems from the recent literature. The main property of the proposed schemes will be tested when we consider numerical experiments featuring stationary solutions. In all test cases, we will consider an ideal gas with $\gamma = 1.4$ and a parameter value $\theta = 1.5$ for the limiter (2.9). The CFL condition is set to 0.485 in (2.2).

2.5.1   Application to the 1D Euler system with gravitational source term

The model

The 1D Euler system with gravitational source term is given by:

$$\begin{cases} \mathbf{u}_t + f(\mathbf{u})_x = S(\mathbf{u}, x), & x \in \Omega \subset \mathbb{R}, \ t > 0. \\ \mathbf{u}(x, 0) = \mathbf{u}_0(x), \end{cases} \tag{2.48}$$

where

$$\mathbf{u} = \begin{pmatrix} \rho \\ \rho u \\ E \end{pmatrix}, f(\mathbf{u}) = \begin{pmatrix} \rho u \\ \rho u^2 + p \\ (E + p)u \end{pmatrix}, S(\mathbf{u}) = \begin{pmatrix} 0 \\ -\rho \phi_x \\ -\rho u \phi_x \end{pmatrix}.$$

Here, ρ is the fluid density, u is the velocity, p is the pressure and $E = \frac{1}{2}\rho u^2 + \frac{p}{\gamma-1}$ is the non-gravitational energy which includes the kinetic and internal energy of the fluid. The gravitational potential $\phi = \phi(x)$ is a given function and γ is the ratio of specific heats. In the absence of the gravitational source term, system (2.48) reduces to a hyperbolic system of conservation laws with a complete set of real eigenvalues and a corresponding set of linearly independent eigenvectors. We present the eigenvalues of the flux jacobian $\frac{\partial f(\mathbf{u})}{\partial \mathbf{u}}$, $\lambda_1 = u$, $\lambda_2 = u + c$, $\lambda_3 = u - c$, where c is the sound speed given by $c = \sqrt{\frac{\gamma p}{\rho}}$.

1D isothermal equilibrium

We start our numerical experiments by verifying that the numerical scheme is capable of preserving any steady state at the discrete level. We consider for our first test case the isothermal equilibrium problem with a linear gravitational field $\phi_x = g = 1$ previously considered in [80]. The numerical solution is computed on 200 grid points of the interval [0,1]. The final time is $t = 0.25$. The equilibrium at the PDE level is defined such that, $\mathbf{u}_t = 0$. The isothermal equilibrium state is given by:

$$\rho(x) = \rho_0 \exp(-\frac{\rho_0 g}{p_0} x),$$
$$u(x) = 0,$$
$$p(x) = p_0 \exp(-\frac{\rho_0 g}{p_0} x).$$

The above formulas for ρ, u, and p ensure that $\mathbf{u}_t = 0$ at the PDE level. However, we need to prove that $\mathbf{u}^{n+1} = \mathbf{u}^n$, in order to prove that the equilibrium is preserved numerically. Here we set $\rho_0 = 1, p_0 = 1$. The reference solution $\tilde{\mathbf{u}}$ chosen in this experiment is exactly the isothermal equilibrium state. The results are illustrated in figures 2.6 and 2.7 where we plot the numerical solution at $t = 0.25$ and we compare it to the exact solution. This figure shows that the equilibrium is exactly preserved and a perfect match between the computed solution and the exact one is observed. Note that in [80], this equilibrium needed a very specific well-balanced strategy to be preserved.

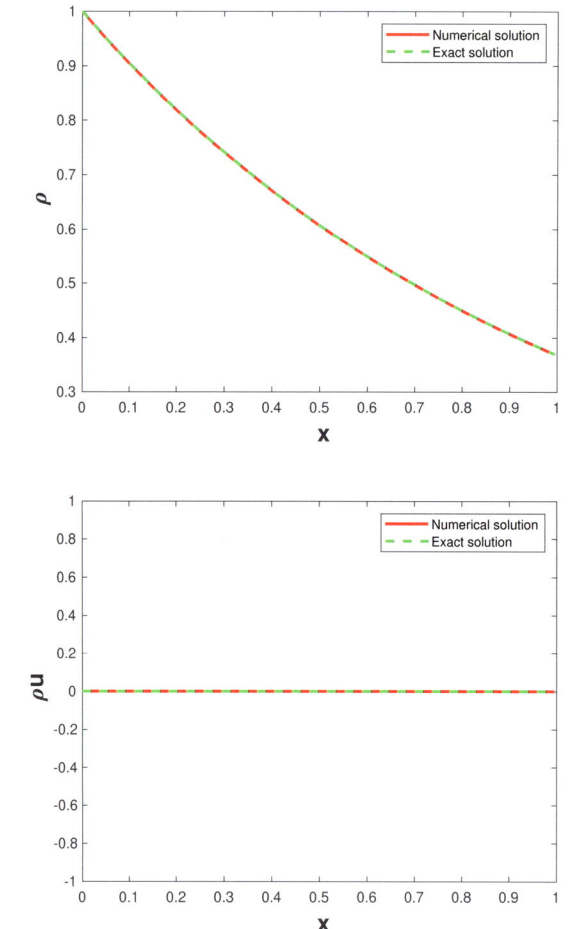

Figure 2.6: 1D isothermal equilibrium: density (top) and momentum (bottom) at time $t = 0.25$.

To test the ability of the scheme to capture perturbations around the equilibrium, a small perturbation is added to the initial pressure. Hence, it is now given as:

$$p(x) = p_0 \exp\left(-\frac{\rho_0 g}{p_0} x\right) + \eta \exp\left(-100 \frac{\rho_0 g}{p_0} (x - 0.5)^2\right),$$

where $\eta = 0.01$. The perturbation will be updated from the pressure at each time by the following formula,

$$k_i^{n+1} = p_i^{n+1} - p_0 \exp\left(-\frac{\rho_0 g}{p_0} x_i\right), \qquad \forall i \text{ and } \forall n.$$

In figure 2.8 we plot the perturbation k obtained at time $t = 0.25$ and we compare it to the initial perturbation on 200 grid points. Outflow boundary conditions are applied. The plots show that the proposed numerical scheme is capable of capturing small perturbations. The order of convergence of the proposed numerical scheme is calculated using the L_1-norm for the density, pressure and the energy components and the obtained results are reported in Table 2.1.

N	L_1-error ρ	Order	L_1-error p	Order	L_1-error E	Order
200	2.765×10^{-6}	—	3.797×10^{-6}	—	9.948×10^{-6}	—
400	7.314×10^{-7}	1.89	1.029×10^{-6}	1.88	2.575×10^{-6}	1.95
800	1.765×10^{-7}	2.05	2.400×10^{-7}	2.10	6.003×10^{-7}	2.19

Table 2.1: 1D isothermal equilibrium: L_1-error and order of convergence.

1D moving equilibrium

Next, we verify that the proposed numerical scheme is capable of preserving moving equilibrium states. We consider the test case previously presented in [82]. A non-linear gravitational field $\phi(x) = \exp(x)(-\exp(x) + \gamma(\exp(-\gamma x))$ is considered. The numerical solution is computed at time $t = 10$ on 200 grid points of the interval [0,1]. The moving equilibrium state is given by:

$$\rho(x) = \rho_0 \exp\left(-\frac{\rho_0 g}{p_0} x\right),$$
$$u(x) = \exp(x),$$
$$p(x) = \exp\left(-\frac{\rho_0 g}{p_0} x\right)^\gamma.$$

$\rho_0 = 1$ and $p_0 = 1$ are given. The considered reference solution in this case is the equilibrium state itself. Figure **??** shows that the density, velocity, energy and pressure are exactly preserved at time $t = 10$. The curves are exactly on top of each other which ensures that the steady state is perfectly preserved with zero error.

1D shock tube problem

We consider for our next experiment the shock tube problem with a linear grav-itational field with $\phi_x = g = 1$, which was previously considered in [80]. The computational domain is the interval $[0,1]$. Reflecting boundary conditions are considered. The reference solution $\tilde{u}$ considered in this experiment is the isothermal equilibrium. Notice here that we are not solving steady state problems, so any other smooth solution could be considered. The initial conditions are given by:

$$\rho(x) = \begin{cases} 1, & \text{if } x \leq 0.5, \\ 0.125, & \text{otherwise,} \end{cases}$$

$$u(x) = 0,$$

$$p(x) = \begin{cases} 1, & \text{if } x \leq 0.5, \\ 0.1, & \text{otherwise.} \end{cases}$$

The numerical solution is computed on 100, 200, and 400 grid points at the final time $t = 0.2$. The obtained results are reported in figures 2.11 and 2.12, where we show the profile of the density, velocity, energy, and pressure. The obtained results are in perfect agreement with those appearing in the literature.

2.5.2 Application to the 2D Euler system with gravitational source term

The model

The 2D Euler system with gravitational source term is given by:

$$\begin{cases} \mathbf{U}_t + F(\mathbf{U})_x + G(\mathbf{U})_y = S(\mathbf{U}), & (x,y) \in \Omega \subset \mathbb{R}^2, \ t > 0. \\ \mathbf{U}(x,y,0) = \mathbf{U}_0(x,y), \end{cases} \tag{2.49}$$

where

$$\mathbf{U} = \begin{pmatrix} \rho \\ \rho u_1 \\ \rho u_2 \\ E \end{pmatrix}, \quad F(\mathbf{U}) = \begin{pmatrix} \rho u_1 \\ \rho u_1^2 + p \\ \rho u_1 u_2 \\ (E+p)u_1 \end{pmatrix}, \quad G(\mathbf{U}) = \begin{pmatrix} \rho u_2 \\ \rho u_1 u_2 \\ \rho u_2^2 + p \\ (E+p)u_2 \end{pmatrix},$$

and

$$S(\mathbf{U}) = \begin{pmatrix} 0 \\ -\rho \phi_x \\ -\rho \phi_y \\ -\rho u_1 \phi_x - \rho u_2 \phi_y \end{pmatrix}.$$

Here ρ is the fluid density, u_1 and u_2 are the velocity components in the x- and y-directions, respectively, p is the pressure and $E = \frac{1}{2}\rho(u_1^2 + u_2^2) + \frac{p}{\gamma-1}$ is the non-gravitational energy which includes the kinetic and internal energy of the fluid. The gravitational potential $\phi = \phi(x,y)$ is a given function and γ is the ratio of specific heats. Similar to the 1D case, and in absence of the gravitational source term, the system reduces to a hyperbolic system of conservation laws with real eigenvalues and a complete set of linearly independent eigenvectors. The eigenvalues of the flux jacobian $\frac{\partial F(U)}{\partial U}$ are, $\lambda_1 = u_1 - c$, $\lambda_2 = u_1$, $\lambda_3 = u_1$, $\lambda_4 = u_1 + c$. For the the flux jacobian $\frac{\partial G(U)}{\partial U}$, $\lambda_1 = u_2 - c$, $\lambda_2 = u_2$, $\lambda_3 = u_2$, $\lambda_4 = u_2 + c$, where c is the sound speed given by $c = \sqrt{\frac{\gamma p}{\rho}}$. In this section we apply the 2D well-balanced unstaggered central scheme we developed in section 2.3 and we solve the classical 2D Euler system with gravitational source term featuring stationary solutions and other equilibrium states.

2D isothermal equilibrium

The first numerical experiment we consider aims to validate the well-balanced property of the proposed 2D scheme. We consider the isothermal equilibrium state problem [19, 82, 80]. This experiment is a direct extension of the 1D experiment previously considered in subsection 2.5.1. The initial conditions correspond to a stationary state and are given by:

$$\rho(x, y) = \rho_0 \exp(-\frac{\rho_0}{p_0}(\phi_x x + \phi_y y)),$$

$$u_1(x, y) = 0,$$

$$u_2(x, y) = 0, \tag{2.50}$$

$$p(x, y) = p_0 \exp(-\frac{\rho_0}{p_0}(\phi_x x + \phi_y y)).$$

$\rho_0 = 1.21$ and $p_0 = 1$ are given. The gravitational potential is linear with $\phi_x = 1$ and $\phi_y = 1$. The computational domain is the square $[0, 1]^2$ discretized using 60×60 grid points. We apply the 2D scheme and compute the numerical solution at the final time $t = 0.25$. Figure 2.13 shows the profile of the density and the energy.

2D Unidirectional equilibrium perturbation

In this test case we extend the 1D perturbation problem to the 2D case where both the equilibrium state and the perturbation are initially set along the x or the y-axis. Whenever set in the x-direction [80], the equilibrium state and the pressure perturbation are given by:

$$\rho(x, y) = \exp(-x)),$$
$$u_1(x, y) = 0,$$
$$u_2(x, y) = 0,$$
$$p(x, y) = \exp(-x) + \eta \exp(-100(x - 0.5)^2).$$

Similar initial data are defined if the perturbation is set in the y-direction. The perturbation will be updated at each time by the following formula,

$$k_{i,j}^n = p_{i,j}^n - \exp(-x_i).$$

The numerical solution is computed at time $t = 0.25$ using our proposed numerical scheme with $\eta = 0.001$. The obtained results are reported in figure 2.15. The observed profiles are similar to those of the 1D case, as well as those reported in the literature. Figure 2.14 shows a comparison between cross sections of the pressure of the 2D problem (with perturbations set in the x- and y- directions) and the corresponding one of the 1D problem. All three curves are in perfect match. The L_1-norm for the density component and the order of convergence of the numerical scheme are reported in table 2.2.

N	L_1-error ρ	Order
200^2	2.8461×10^{-7}	—
400^2	7.0611×10^{-8}	2.01
800^2	1.6840×10^{-8}	2.06

Table 2.2: 2D Unidirectional equilibrium perturbation: L_1-error and order of convergence for the density.

2D moving equilibrium

This test case is an extension of the 1D moving equilibrium problem to the 2D case; it is meant to verify that the proposed numerical scheme is capable of preserving 2D steady states with non-zero velocities. The initial coefficients are given by:

$$\rho(x, y) = \rho_0 \exp(-\frac{\rho_0 g}{p_0}(x + y)),$$
$$u_1(x, y) = \exp(x + y),$$
$$u_2(x, y) = \exp(x + y),$$
$$p(x, y) = \exp(-\frac{\rho_0 g}{p_0}(x + y))^\gamma.$$

$\rho_0 = 1$, $p_0 = 1$, and $g = 1$. We consider a nonlinear gravitational potential given by $\phi(x, y) = \exp(x + y)(-\exp(x + y) + \gamma(\exp(-\gamma(x + y))))$. The numerical solution is computed at the final time $t = 0.25$. The equilibrium is preserved exactly and a 1D/2D comparison is held on the density component at the final time in figure 2.16. The comparison shows a perfect match, thus confirming the potential of the proposed scheme to handle stationary equilibria.

2D shock tube problem

We consider for our last experiment the 2D sod shock tube problem. As in the 1D case, the reference solution $\tilde{U}$ is the isothermal equilibrium solution (2.50).

We consider first the flow along the x-direction with the linear gravitational field with $\phi_x = 1$ and $\phi_y = 0$; the initial data are given by:

$$\rho(x, y) = \begin{cases} 1, & \text{if } x \leq 0.5, \\ 0.125, & \text{otherwise.} \end{cases}$$

$$u_1(x, y) = 0 = u_2(x, y).$$

$$p(x, y) = \begin{cases} 1, & \text{if } x \leq 0.5, \\ 0.1, & \text{otherwise.} \end{cases}$$

The computational domain is the square $[0, 1]^2$ discretized using 400×10 grid points. In a similar way, we define the initial data along the y-direction, where the same computational domain is discretized using 10×400 grid points. The numerical solution is computed at the final time $t = 0.2$ using the proposed well-balanced scheme. The obtained numerical results are reported in figures 2.17 and 2.18 where we present a comparison between cross sections of the 2D problem set along the x- and y- directions for the density, velocity, energy and pressure and the corresponding solution of the 1D problem. A perfect match between the plots is observed and the obtained results are in perfect agreement with corresponding ones appearing in the literature.

2.5.3 Application to the 2D MHD equations with gravitational source term

The model

Ideal Magnetohydrodynamics (MHD) equations model problems in physics and astrophysics. The MHD system is a combination of the Navier-Stokes equations of fluid dynamics and the Maxwell equations of electromagnetism. A gravitational source term is added to the ideal MHD equations in two space dimensions in order to model more complicated problems arising in astrophysics and solar physics such as modeling wave propagation in idealized stellar atmospheres [69, 11]. The system of MHD equations with gravitational source term in two space dimensions is given by:

$$\begin{cases} \mathbf{U}_t + F(\mathbf{U})_x + G(\mathbf{U})_y = S(\mathbf{U}), & (x, y) \in \Omega \subset \mathbb{R}^2, \ t > 0. \\ \mathbf{U}(x, y, 0) = \mathbf{U}_0(x, y), \end{cases} \tag{2.51}$$

where

$$\mathbf{U} = \begin{pmatrix} \rho \\ \rho u_1 \\ \rho u_2 \\ \rho u_3 \\ E \\ B_1 \\ B_2 \\ B_3 \end{pmatrix}, \ F(\mathbf{U}) = \begin{pmatrix} \rho u_1 \\ \rho u_1^2 + \Pi_{11} \\ \rho u_1 u_2 + \Pi_{12} \\ \rho u_1 u_3 + \Pi_{13} \\ E u_1 + u_1 \Pi_{11} + u_2 \Pi_{12} + u_3 \Pi_{13} \\ 0 \\ \Lambda_2 \\ -\Lambda_3 \end{pmatrix},$$

$$G(\mathbf{U}) = \begin{pmatrix} \rho u_2 \\ \rho u_2 u_1 + \Pi_{21} \\ \rho u_2^2 + \Pi_{22} \\ \rho u_2 u_3 + \Pi_{23} \\ E u_2 + u_1 \Pi_{21} + u_2 \Pi_{22} + u_3 \Pi_{23} \\ -\Lambda_3 \\ 0 \\ \Lambda_1 \end{pmatrix}, S(\mathbf{U}) = \begin{pmatrix} 0 \\ 0 \\ -\rho \phi_y \\ 0 \\ -\rho u_2 \phi_y \\ 0 \\ 0 \\ 0 \end{pmatrix}.$$

Here ρ is the fluid density, $\rho \mathbf{u}$ is the momentum with $\mathbf{u} = (u_1, u_2, u_3)$, p is the pressure, $\mathbf{B} = (B_1, B_2, B_3)$ is the magnetic field, and E is the kinetic and internal energy of the fluid given by the following equation $E = \frac{p}{\gamma - 1} + \frac{1}{2}\rho|\mathbf{u}|^2 + \frac{1}{2}|\mathbf{B}|^2$ with γ the ratio of specific heats. $\phi = \phi(x, y)$, with $\phi_x = 0$ and $\phi_y = g$, is the gravitational potential and it is a given function. $\Lambda = \mathbf{u} \times \mathbf{B}$, Π_{11}, Π_{22} and Π_{33} are the diagonal elements of the total pressure tensor and Π_{12}, Π_{13} and Π_{23} are the off-diagonal tensor are given by the following formulas:

$\Pi_{ii} = p + \frac{1}{2}(B_j^2 + B_k^2 - B_i^2)$ and $\Pi_{ij} = -\frac{1}{2}B_i B_j$, for $i, j, k = 1, 2, 3$.

To determine the time-step using the CFL condition (2.2), we present the eigenvalues of the flux jacobian in the x-direction,

$\lambda_1 = u_1 - c_f, \lambda_2 = u_1 - b_1, \lambda_3 = u_1 - c_s, \lambda_4 = u_1, \lambda_5 = u_1, \lambda_6 = u_1 + c_s, \lambda_7 = u_1 + b_1, \lambda_8 = u_1 + c_f$. The eigenvalues of the flux jacobian in the y-direction are analogously defined.

Here,

$$c_f = \sqrt{\frac{1}{2}\left(a^2 + b^2 + \sqrt{(a^2 + b^2)^2 - 4a^2 b_1^2}\right)}, \tag{2.52}$$

and

$$c_s = \sqrt{\frac{1}{2}\left(a^2 + b^2 - \sqrt{(a^2 + b^2)^2 - 4a^2 b_1^2}\right)}, \tag{2.53}$$

are respectively the fast and slow wave speed with $a = \sqrt{\frac{\gamma p}{\rho}}$ is the sound speed and $b = \sqrt{b_1^2 + b_2^2 + b_3^2}$ with $b_i = \frac{B_i}{\sqrt{\rho}}, i \in \{1, 2, 3\}$. For additional reading on the hyperbolic analysis of the system, readers are refered to [32, 64].

The conservation of momentum is exposed to Lorentz force from the magnetic field and to gravitational force. In addition, the conservation of the total energy (internal, kinetic and magnetic) has the gravitational potential energy as a source term. A list of numerical experiments has been considered in order to verify the robustness and accuracy of our method in the case of the system of MHD equations.

Constrained Transport Method (CTM)

From electromagnetic theory, the magnetic field $\mathbf{B}$ must be solenoidal i.e. $\nabla \cdot \mathbf{B} = 0$ at all times. The divergence-free constraint on the magnetic field reflects the fact that magnetic mono-poles have not been observed in nature. The induction equation for updating the magnetic field imposes the divergence on the magnetic field. Hence, a numerical scheme for the MHD equations should maintain the divergence-free property of the discrete magnetic field at each time-step. Numerical schemes usually fail to satisfy the divergence-free constraint and numerical instabilities and unphysical oscillations may be observed [72]. Several methods were developed to overcome this issue. The projection method, in which the magnetic field is projected into a zero divergence field by solving an elliptic equation at each time step [13]. Another procedure is the Godunov-Powell procedure [65, 66, 31], where the Godunov-Powell form of the system of the MHD equations is discretized instead of the original system. The Godunov-Powell system has the divergence of the magnetic field as a part of the source term. Hence, divergence errors are transported out of the domain with the flow. A third approach is the CTM [14, 67, 28]. The CTM was modified from its original form to the case of staggered central schemes [3]. It was later extended to the case of unstaggered central schemes [75]. In this work we consider the version of CTM developed in [75]. At the end of each iteration, we apply the CTM corrections to the magnetic field components. Starting from a magnetic field that satisfies the divergence-free constraint $\nabla \cdot \mathbf{B}_{i,j}^{n} = 0$, we would like to prove $\nabla \cdot \mathbf{B}_{i,j}^{n+1} = 0$. The discrete divergence using central differences at time t^{n} is given by,

$$
\begin{aligned}
\nabla \cdot \mathbf{B}_{i,j}^{n} &= \left(\frac{\partial B_x}{\partial x}\right)_{i,j}^{n} + \left(\frac{\partial B_y}{\partial y}\right)_{i,j}^{n} \\
&= \frac{(B_x)_{i+1,j}^{n} - (B_x)_{i-1,j}^{n}}{2\Delta x} + \frac{(B_y)_{i,j+1}^{n} - (B_y)_{i,j-1}^{n}}{2\Delta y} \\
&= 0.
\end{aligned}
$$

The vector of conserved variables $\mathbf{U}^{n+1}$ is computed by the numerical scheme, but $\nabla.\mathbf{B}_{i,j}^{n+1}$ might not be zero. Therefore, we compute the magnetic field $\mathbf{B}_{i,j}^{n+1}$ by discretizing the induction equation at the cell centers of $C_{i,j}$,

$$
\frac{\partial}{\partial t}\begin{pmatrix} B_x \\ B_y \end{pmatrix} - \frac{\partial}{\partial x}\begin{pmatrix} 0 \\ \Omega \end{pmatrix} + \frac{\partial}{\partial y}\begin{pmatrix} \Omega \\ 0 \end{pmatrix} = 0,
$$

where $\Omega = (-\mathbf{u} \times \mathbf{B})_z = -u_x B_y + u_y B_x$. Hence, the discretization of the induction equation is the following,

$$
\begin{cases}
\dfrac{(B_x)_{i+\frac{1}{2},j+\frac{1}{2}}^{n+1} - (B_x)_{i+\frac{1}{2},j+\frac{1}{2}}^{n}}{\Delta t} + \dfrac{\Omega_{i+\frac{1}{2},j+\frac{3}{2}}^{n+\frac{1}{2}} - \Omega_{i+\frac{1}{2},j-\frac{1}{2}}^{n+\frac{1}{2}}}{2\Delta y} = 0, \\[3mm]
\dfrac{(B_y)_{i+\frac{1}{2},j+\frac{1}{2}}^{n+1} - (B_y)_{i+\frac{1}{2},j+\frac{1}{2}}^{n}}{\Delta t} - \dfrac{\Omega_{i+\frac{3}{2},j+\frac{1}{2}}^{n+\frac{1}{2}} - \Omega_{i-\frac{1}{2},j+\frac{1}{2}}^{n+\frac{1}{2}}}{2\Delta x} = 0.
\end{cases}
$$

Then,

$$
\begin{cases}
(B_x)_{i+\frac{1}{2},j+\frac{1}{2}}^{n+1} = (B_x)_{i+\frac{1}{2},j+\frac{1}{2}}^{n} - \dfrac{\Delta t}{2\Delta y}\left(\Omega_{i+\frac{1}{2},j+\frac{3}{2}}^{n+\frac{1}{2}} - \Omega_{i+\frac{1}{2},j-\frac{1}{2}}^{n+\frac{1}{2}}\right), \\[3mm]
(B_y)_{i+\frac{1}{2},j+\frac{1}{2}}^{n+1} = (B_y)_{i+\frac{1}{2},j+\frac{1}{2}}^{n} + \dfrac{\Delta t}{2\Delta x}\left(\Omega_{i+\frac{3}{2},j+\frac{1}{2}}^{n+\frac{1}{2}} - \Omega_{i-\frac{1}{2},j+\frac{1}{2}}^{n+\frac{1}{2}}\right).
\end{cases}
\tag{2.54}
$$

Now, we compute $\Omega_{i+\frac{1}{2},j+\frac{1}{2}}^{n+\frac{1}{2}}$ using the numerical solution computed at time t^n and t^{n+1} in order to obtain second order of accuracy in time,

$$
\begin{aligned}
\Omega_{i+\frac{1}{2},j+\frac{1}{2}}^{n+\frac{1}{2}} &= \frac{1}{2}\left[\Omega_{i+\frac{1}{2},j+\frac{1}{2}}^{n+1} + \Omega_{i+\frac{1}{2},j+\frac{1}{2}}^{n}\right], \\
&= \frac{1}{2}\left[\Omega_{i+\frac{1}{2},j+\frac{1}{2}}^{n+1} + \frac{\Omega_{i,j}^{n} + \Omega_{i+1,j}^{n} + \Omega_{i,j+1}^{n} + \Omega_{i+1,j+1}^{n}}{4}\right].
\end{aligned}
$$

Next, we calculate $\nabla.(\mathbf{B})_{i+\frac{1}{2},j+\frac{1}{2}}^{n+1}$

$$
\nabla.(\mathbf{B})_{i+\frac{1}{2},j+\frac{1}{2}}^{n+1} = \underbrace{\frac{(B_x)_{i+\frac{3}{2},j+\frac{1}{2}}^{n+1} - (B_x)_{i-\frac{1}{2},j+\frac{1}{2}}^{n+1}}{2\Delta x}}_{=\text{I}} + \underbrace{\frac{(B_y)_{i+\frac{1}{2},j+\frac{3}{2}}^{n+1} - (B_y)_{i+\frac{1}{2},j-\frac{1}{2}}^{n+1}}{2\Delta y}}_{=\text{J}}.
$$

$$(2.55)$$

We compute now I and J as

$$
\begin{aligned}
I &= \frac{(B_x)_{i+\frac{3}{2},j+\frac{1}{2}}^{n+1} - (B_x)_{i-\frac{1}{2},j+\frac{1}{2}}^{n+1}}{2\Delta x}, \\
&= \frac{1}{2\Delta x}\left[\frac{(B_x)_{i+1,j}^{n} + (B_x)_{i+2,j+1}^{n} + (B_x)_{i+2,j}^{n} + (B_x)_{i+1,j+1}^{n}}{4}\right. \\
&\quad - \frac{\Delta t}{2\Delta y}\left(\Omega_{i+\frac{3}{2},j+\frac{3}{2}}^{n+\frac{1}{2}} - \Omega_{i+\frac{3}{2},j-\frac{1}{2}}^{n+\frac{1}{2}}\right) - \frac{(B_x)_{i-1,j}^{n} + (B_x)_{i,j+1}^{n} + (B_x)_{i,j}^{n} + (B_x)_{i-1,j+1}^{n}}{4} \\
&\quad \left. + \frac{\Delta t}{2\Delta y}\left(\Omega_{i-\frac{1}{2},j+\frac{3}{2}}^{n+\frac{1}{2}} - \Omega_{i-\frac{1}{2},j-\frac{1}{2}}^{n+\frac{1}{2}}\right)\right].
\end{aligned}
$$

$$
\begin{aligned}
J &= \frac{(B_y)_{i+\frac{1}{2},j+\frac{3}{2}}^{n+1} - (B_y)_{i+\frac{1}{2},j-\frac{1}{2}}^{n+1}}{2\Delta y}, \\
&= \frac{1}{2\Delta y}\left[\frac{(B_y)_{i,j+1}^{n} + (B_y)_{i+1,j+1}^{n} + (B_y)_{i+1,j+2}^{n} + (B_y)_{i,j+2}^{n}}{4}\right. \\
&\quad + \frac{\Delta t}{2\Delta x}\left(\Omega_{i+\frac{3}{2},j+\frac{3}{2}}^{n+\frac{1}{2}} - \Omega_{i-\frac{1}{2},j+\frac{3}{2}}^{n+\frac{1}{2}}\right) - \frac{(B_y)_{i,j-1}^{n} + (B_y)_{i+1,j-1}^{n} + (B_y)_{i+1,j}^{n} + (B_y)_{i,j}^{n}}{4} \\
&\quad \left. - \frac{\Delta t}{2\Delta x}\left(\Omega_{i+\frac{3}{2},j-\frac{1}{2}}^{n+\frac{1}{2}} - \Omega_{i-\frac{1}{2},j-\frac{1}{2}}^{n+\frac{1}{2}}\right)\right].
\end{aligned}
$$

The sum of I and J is,

$$
\begin{aligned}
I + J = \frac{1}{8\Delta x} \Big[& (B_x)_{i+1,j}^n - (B_x)_{i-1,j}^n + (B_x)_{i+2,j+1}^n - (B_x)_{i,j+1}^n \\
& + (B_x)_{i+2,j}^n - (B_x)_{i,j}^n + (B_x)_{i+1,j+1}^n - (B_x)_{i-1,j+1}^n \Big] \\
+ \frac{\Delta t}{4\Delta x \Delta y} \Big[& \Big(-\Omega_{i+\frac{3}{2},j+\frac{3}{2}}^{n+\frac{1}{2}} + \Omega_{i+\frac{3}{2},j+\frac{3}{2}}^{n+\frac{1}{2}} \Big) + \Big(\Omega_{i+\frac{3}{2},j-\frac{1}{2}}^{n+\frac{1}{2}} - \Omega_{i+\frac{3}{2},j-\frac{1}{2}}^{n+\frac{1}{2}} \Big) \\
& + \Big(\Omega_{i-\frac{1}{2},j+\frac{3}{2}}^{n+\frac{1}{2}} - \Omega_{i-\frac{3}{2},j+\frac{3}{2}}^{n+\frac{1}{2}} \Big) + \Big(\Omega_{i-\frac{1}{2},j-\frac{1}{2}}^{n+\frac{1}{2}} - \Omega_{i-\frac{3}{2},j-\frac{1}{2}}^{n+\frac{1}{2}} \Big) \Big] \\
+ \frac{1}{8\Delta y} \Big[& (B_y)_{i,j+1}^n - (B_y)_{i,j-1}^n + (B_y)_{i+1,j+1}^n - (B_y)_{i+1,j-1}^n \\
& + (B_y)_{i+1,j+2}^n - (B_y)_{i+1,j}^n + (B_y)_{i,j+2}^n - (B_y)_{i,j}^n \Big].
\end{aligned}
$$

Hence,

$$
\begin{aligned}
I + J = \frac{1}{4} \Bigg[& \frac{(B_x)_{i+1,j}^n - (B_x)_{i-1,j}^n}{2\Delta x} + \frac{(B_y)_{i,j+1}^n - (B_y)_{i,j-1}^n}{2\Delta y} \\
& + \frac{(B_x)_{i+2,j+1}^n - (B_x)_{i,j+1}^n}{2\Delta x} + \frac{(B_y)_{i+1,j+2}^n - (B_y)_{i+1,j}^n}{2\Delta y} \\
& + \frac{(B_x)_{i+2,j}^n - (B_x)_{i,j}^n}{2\Delta x} + \frac{(B_y)_{i,j+2}^n - (B_y)_{i,j}^n}{2\Delta y} \\
& + \frac{(B_x)_{i+1,j+1}^n - (B_x)_{i-1,j+1}^n}{2\Delta x} + \frac{(B_y)_{i+1,j+1}^n - (B_y)_{i+1,j-1}^n}{2\Delta y} \Bigg],
\end{aligned} \quad (2.56)
$$

and the divergence of the magnetic field on the staggered grid $\nabla \cdot (\mathbf{B})_{i+\frac{1}{2},j+\frac{1}{2}}^{n+1}$ reduces to,

$$
\nabla \cdot (\mathbf{B})_{i+\frac{1}{2},j+\frac{1}{2}}^{n+1} = \frac{1}{4} \left[\nabla \cdot \mathbf{B}_{i,j}^n + \nabla \cdot \mathbf{B}_{i+1,j+1}^n + \nabla \cdot \mathbf{B}_{i+1,j}^n + \nabla \cdot \mathbf{B}_{i,j+1}^n \right] = 0. \quad (2.57)
$$

Finally, we compute the magnetic field on the main grid $\mathbf{B}_{i,j}^{n+1}$ as the average of its values on the staggered grid,

$$
\mathbf{B}_{i,j}^{n+1} = \frac{1}{4} \left[\mathbf{B}_{i+\frac{1}{2},j+\frac{1}{2}}^{n+1} + \mathbf{B}_{i+\frac{1}{2},j-\frac{1}{2}}^{n+1} + \mathbf{B}_{i-\frac{1}{2},j+\frac{1}{2}}^{n+1} + \mathbf{B}_{i-\frac{1}{2},j-\frac{1}{2}}^{n+1} \right].
$$

Hence,

$$
\nabla \cdot \mathbf{B}_{i,j}^{n+1} = 0. \quad (2.58)
$$

On a side note, the CTM maintains the second order of the base scheme as discretizations were performed with second order of accuracy.

Smooth solution

In order to numerically validate the second order accuracy of our scheme, we consider the first test case from [83]. For simplcity, we consider a variant of the the flow presented in [83]. The initial data is an MHD sine wave propagating over the computational domain [-2,2]×[-2,2] until time $t = 0.1$,

$$U = [\rho, u_1, u_2, u_3, B_1, B_2, B_3, p] = [1 + 0.99 \sin(x - 2t), 1, 1, 0, 0.1, 0.1, 0, 1], \quad (2.59)$$

with $\gamma = 1.4$. Tabel 2.3 lists the numerical L_1-errors for the density component on different grids together with the order of the scheme.

$N \times N$	L_1-error ρ	Order
200×200	8.0190×10^{-4}	—
400×400	1.9357×10^{-4}	2.05
800×800	4.5111×10^{-5}	2.10
1600×1600	9.6607×10^{-6}	2.22

Table 2.3: Smooth solution: L_1-error and order of convergence.

2D shock tube problem

For the first numerical test case, we consider a shock tube problem for the system of ideal MHD equations extracted from [4]. The simulation takes place over the computational domain $[-1, 1] \times [0, 1]$. $U = [\rho, u_1, u_2, u_3, B_2, B_3, p]$ is initially given as $U = [1, 0, 0, 0, \sqrt{4}, 0, 1]$ for $x < 0.5$ and $U = [0.125, 0, 0, 0, -\sqrt{4}, 0, 0.1]$ for $x > 0.5$ and $B_1 = 0.75\sqrt{4}$. This test case features seven discontinuities. It was originally introduced for the non-scaled MHD equations [4]. Hence, removing π from the initial data makes it a valid test case for the scaled MHD equations. We compute the solution at the final time $t = 0.25$ on 400×400 grid. Because the numerical divergence at the final time was zero, there was no need to apply the CTM. The cross sections in figures 2.19 and 2.20 show a very good agreement with the results in the literature. In order to investigate the effect of the CTM on the computed solution, we did a convergence study in figure 2.21 and (2.22) while applying the CTM. As it is very clear in the figures above, applying the CTM for the UC schemes has a small smearing out effect on the solution.

Four stages Ideal MHD Riemann problem

This test case is considered to prove the ability of our scheme to solve ideal MHD problems and preserve the divergence-free constraint. The initial data consist of four constant states [4, 75] . The initial four constant states are given as follows,

$$(\rho, u_1, u_2, p) = \begin{cases} (1, 0.75, 0.5, 1) & \text{if } x > 0 \text{ and } y > 0 \\ (2, 0.75, 0.5, 1) & \text{if } x < 0 \text{ and } y > 0 \\ (1, -0.75, 0.5, 1) & \text{if } x < 0 \text{ and } y < 0 \\ (3, -0.75, -0.5, 1) & \text{if } x > 0 \text{ and } y < 0 \end{cases} \quad (2.60)$$

with an initial uniform magnetic field $\mathbf{B} = (2,0,1)$. The numerical solution is computed in the square $[-1,1] \times [-1,1]$ on 400×400 grid points.

Figure 2.23 illustrates the density at the final time $t_f = 0.8$ with and without applying constrained transport treatment to the magnetic field components. Similar comparison on the divergence of the magnetic field is illustrated in figure 2.24. The results highlight the robustness of the numerical scheme in the sense that even without treatment we are able to show numerical simulation while other schemes simply blow up without special treatment of the magnetic field.

MHD vortex

For our third test case, we consider the MHD vortex for the homogeneous ideal MHD equations [8]. The initial data represent a moving stationary solution of the system of the ideal MHD equations and are given by, $r^2 = x^2 + y^2$, $\rho = 1$, $u_1 = u_0 - \kappa_p \exp(\frac{1-r^2}{2})y$, $u_2 = v_0 + \kappa_p \exp(\frac{1-r^2}{2})x$, $u_3 = 0$, $B_1 = -m_p \exp(\frac{1-r^2}{2})y$, $B_2 = -m_p \exp(\frac{1-r^2}{2})x$, $B_3 = 0$, and $p = 1 + \left(\frac{m_p^2}{2}(1 - r^2) - \frac{\kappa_p^2}{2} \right)$. We set the parameters $m_p = 1, \kappa_p = 1, u_0 = 0$, and $v_0 = 0$. The vortex is advected through the domain $[-5,5] \times [-5,5]$ with a velocity (u_0, v_0). Steady state boundary conditions are used in this test case. In figure 2.25, we present the pressure profile at the final time $t = 100 \frac{2\pi}{\sqrt{e}\kappa_p} \approx 100 \frac{3.14}{\kappa_p}$ on different grids. The steady state gets preserved exactly as the background solution $\ddot{\mathbf{U}}$ is the vortex itself. No treatment of the magnetic field has been done in [8] as the selection of the MHD test cases is restricted to this steady state, which gets preserved exactly. However, treating the magnetic field is a necessity in other test cases as we will see later.

Hydrodynamic wave propagation

The aim of this test case is to test the well-balanced property of the subtraction method by simulating a steady state solution under hydrodynamic wave propagation. The experiment is carried out in two steps. The first step is to check that the subtraction method preserves the steady state. The initial data are the hydrodynamic steady state in the computational domain $[0,4] \times [0,1]$.

$$\rho(x,y) = \rho_0 \exp(-\frac{y}{H}), p(x,y) = p_0 \exp(-\frac{y}{H}), \mathbf{u} = 0, \mathbf{B} = 0. \tag{2.61}$$

With $H = \frac{p_0}{g\rho_0} = 0.158$, $p_0 = 1.13$ and $g = 2.74$. The subtraction method preserves the hydrodynamic steady state exactly after choosing the reference solution $\tilde{\mathbf{U}}$ at the steady state itself. Figure 2.26 shows a very simple comparison of the density and the energy cross section at $t = 0$ and the final time $t = 1.8$. The second step is to add perturbation to the steady state as a time dependent sinusoidal wave that propagates from the bottom boundary of the vertical velocity and exits from the top one. The wave formula is as the following,

$$u^n_{2i,\{0,-1\}} = \exp(-100(x_{i,\{0,-1\}} - 1.9)^2)c\sin(6\pi t^n). \tag{2.62}$$

The bottom boundary is a localized piston at $x = 1.9$. Figure 2.27 shows the profile of the wave at the final time $t = 1.8$ for $c = 0.003$ (left) and for $c = 0.3$ (right) for 800×200 grid points. The waves propagate in both cases from bottom to top under the effect of the pressure and gravity forces. The case where $c = 0.003$ models a small perturbation and $c = 0.3$ models a stronger wave. The results are in a very good agreement with the ones in [31]. More importatntly they match the results of the most accurate (third order) of the three schemes compared in [31]. Hence, the scheme is well-balanced in the sense that it preserves the steady state and can capture its perturbations.

MHD wave propagation

In this test case, we model propagating waves that not only undergo the effects of pressure and gravity, but also that of the magnetic field. The test case is extracted from [31]. We consider the magnetohydrodynamic steady state defined as,

$$\rho(x,y) = \rho_0 \exp(-\frac{y}{H}), p(x,y) = p_0 \exp(-\frac{y}{H}), \mathbf{u} = 0, \mathbf{B} = (0, \mu, 0), \nabla \cdot \mathbf{B} = 0.$$
(2.63)

Where μ is a parameter that takes different values for each part of the experiment. The waves model a perturbation of the steady state that starts from the bottom boundary of the normal velocity as the following,

$$\mathbf{u}_{i,\{0,1\}}^n = \begin{cases} \frac{\mathbf{B}_{i,\{0,1\}}}{|\mathbf{B}_{i,\{0,1\}}|} c \sin(6\pi t^n) & \text{for } x \in [0.95, 1.05], \\ 0 & \text{Otherwise,} \end{cases}$$
(2.64)

with $c = 0.3$. The computational domain is $[0, 2] \times [0, 1]$. We use the wave propagation boundary conditions suggested in [31]. These boundaries are periodic boundaries in the x-direction for $\mathbf{U}$ and p and Neumann type boundary conditions in the y-direction as the following,

$$\rho_{i,1}^n = \rho_{i,2}^n e^{\frac{\Delta y}{H}}, \rho_{i,0}^n = \rho_{i,1}^n e^{\frac{\Delta y}{H}}$$
$$\rho_{i,ny-1}^n = \rho_{i,ny-2}^n e^{\frac{-\Delta y}{H}}, \rho_{i,ny}^n = \rho_{i,ny-1}^n e^{\frac{-\Delta y}{H}}$$

for $1 \leq i \leq nx$. Similar boundaries for the momentum $\rho\mathbf{u}$ and the pressure p. Energy boundaries are computed from the pressure. For the magnetic field boundaries, we simply copy the data from the cell before. We present the profile of the velocity in the direction of the magnetic field,

$$u_B = <\mathbf{u}, \mathbf{B}> /|\mathbf{B}|,$$
(2.65)

at the final time $t = 0.54$ for different values of μ. As μ increases, the effect of the magnetic field on the propagating wave increases. The wave profile gets compressed as the magnetic field takes higher values. The plasma parameter is given by $\beta = \frac{2p}{\mathbf{B}^2}$

[31]. It measures the relative strength of the thermal pressure to the magnetic field, and is crucial in determining the dynamics of the plasma. The β-isolines are illustrated in black and the lines of the magnetic field are illustrated in white. The parameter β indicates the effects of the pressure and the magnetic field on the propagating wave such that, for $\beta > 1$, the region is pressure dominated, while for $\beta < 1$, the region is magnetic field dominated. In figure 2.28, the profile of the velocity in the direction of the magnetic field, in the case of μ almost zero, is illustrated, which is exactly the velocity in the y-direction in this case. The wave propagates freely along the computational domain taking a radial profile in the absence of the magnetic field on 400×200 grid points. Figure 2.29, shows the profile of the propgating wave under the effect of a stronger magnetic field for $\mu = 1$ on 400×200 grid points without applying CTM. In addition, figure 2.29 presents the divergence of the magnetic field which is clearly not zero. On the other hand, we present the same results with applying CTM on 1200×600 grid points in figure 2.30. Applying the CTM results in a zero discrete divergence of the magnetic field up to machine precision. Another effect of applying the CTM is the diffusion we see in figure 2.30, which was resolved by evolving the solution on a finer grid. Additionally, we present the velocity in the direction perpendicular to the magnetic field in figure 2.31 for $\mu = 1$ at different times.

Our results, obtained with the second order scheme, are comparable with the results in [31], obtained with third order schemes, which ensures the robustness of our scheme and its capability of solving physically challenging problems, such as wave propagation under the effect of pressure and gravity.

2.5.4   MHD wave propagation - weak magnetic field

In order to test the effect of a more complicated non-constant magnetic field on the wave propagation, we choose a test case featuring a non-trivial magnetic steady state from [31],

$$\rho(x,y) = \rho_0 \exp(-\frac{y}{H}), \; p(x,y) = p_0 \exp(-\frac{y}{H}), \; \mathbf{u} = 0, \; \nabla \cdot \mathbf{B} = 0, \tag{2.66}$$

where $\mathbf{B}$ is defined as Fourier expansion of vector harmonic functions,

$$\mathbf{B} = (B_1, \; B_2, \; B_3) = \left(\sum_{k=0}^{M} f_k \sin\left(\frac{2k\pi x}{X}\right) e^{-\frac{2k\pi x}{X}}, \; \sum_{k=0}^{M} f_k \cos\left(\frac{2k\pi x}{X}\right) e^{-\frac{2k\pi x}{X}}, \; 0 \right)$$
$$\tag{2.67}$$

with f_k are the Fourier coefficients given by the vector $\mathbf{FR}/3$ where $\mathbf{FR}$ is defined as,

$$\begin{aligned} \mathbf{FR} = \{f_0, ..., f_{14}\} = \{ & 0.552802906842; \; -0.696736253842; \; 0.908809914778; \\ & -0.813921192337; \; 0.360524088458; \; 0.115217242296; \; -0.281974513346; \\ & 0.143723957761; \; 0.049431756210; \; -0.110095259045; \; 0.053464228949; \\ & 0.011695376102; \; -0.028284735991; \; 0.013116555865; \; 0.001434008866 \; \}. \end{aligned} \tag{2.68}$$

M is the total number of Fourier modes, which is 14 in this case. X equals 4 is the length of the domain in the x-direction. The simulation is carried in the computational domain $[0, 4] \times [0, 1]$. After preserving the steady state, perturbation (2.64) is added to the velocity in the y-direction with magnitude $c = 3 \times 10^{-4}$. The results on a 800×200 mesh at the final time $t = 0.9$ is illustrated in figure (2.32). This test case shows the ability of the scheme to capture very small perturbations around the steady state. This was possible because of the fact that our scheme preserves the steady state exactly. Figure (2.33) shows the discrete divergence of the magnetic field initially and at the final time. In this case, we do not need to apply the CTM as the initial discrete divergence is preserved.

2.6 Conclusion

In conclusion, we develop 1D and 2D second order unstaggered finite volume central schemes for general balance laws. The proposed scheme is capable of preserving any type of known equilibrium states due to a special reformulation that computes the numerical solution in terms of a specific reference state. Applications to the systems of Euler and MHD equations with gravitational source term are presented in the numerical results section. A comparison between the obtained numerical results and the corresponding literature ensures the robustness and the accuracy of the developed schemes. In this work, we chose the CTM as a procedure to clean the divergence of the magnetic field. We realized that, it has a smearing out effect on the solution especially in the physically challenging test cases. For this reason, the CTM is applied dynamically whenever needed. Meaning that, in the test cases where the numerical divergence is zero at the final time and no numerical instabilities has been observed, we do not apply it. This leaves us with a second order well-balanced finite volume numerical scheme that captures solutions of the MHD equations and satisfies the divergence-free constraint.

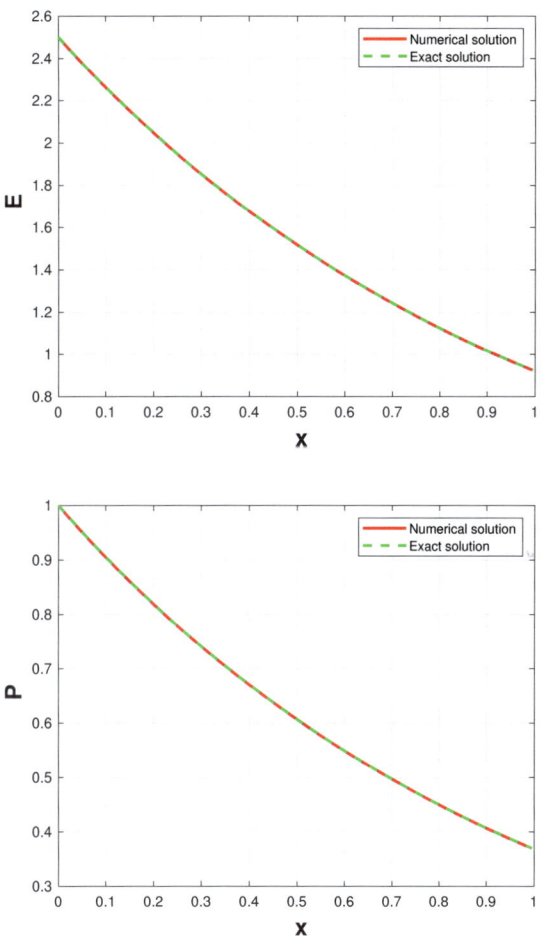

Figure 2.7: 1D isothermal equilibrium: energy (top) and pressure (bottom) at time $t = 0.25$.

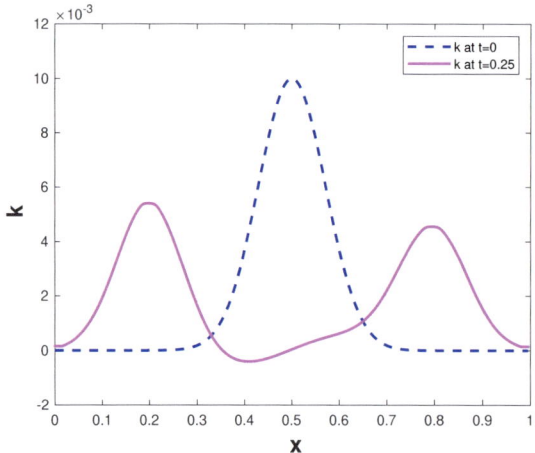

Figure 2.8: 1D isothermal equilibrium: profile of the initial perturbation (dashed curve) and the perturbation at the final time $t = 0.25$ (dotted curve).

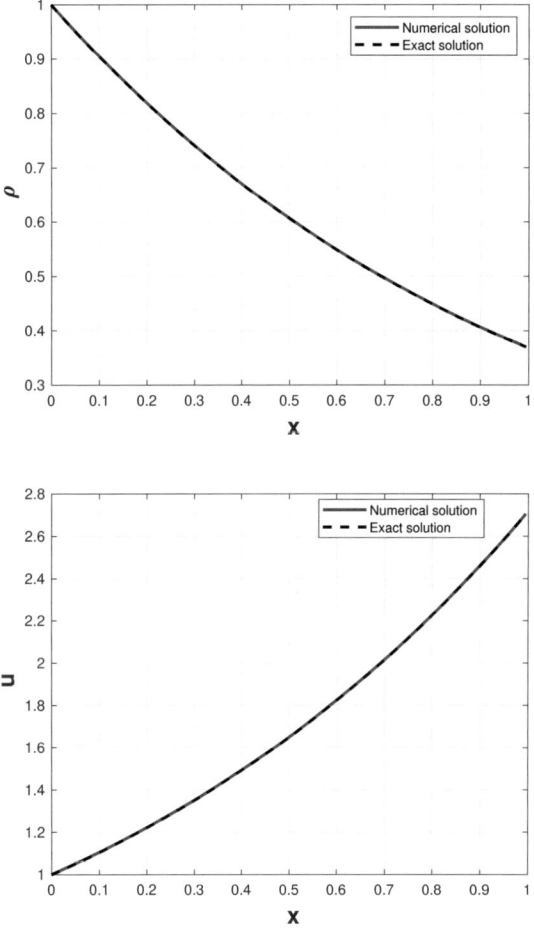

Figure 2.9: 1D moving equilibrium: profile of the density (top) and velocity (bottom) obtained at time $t = 10$.

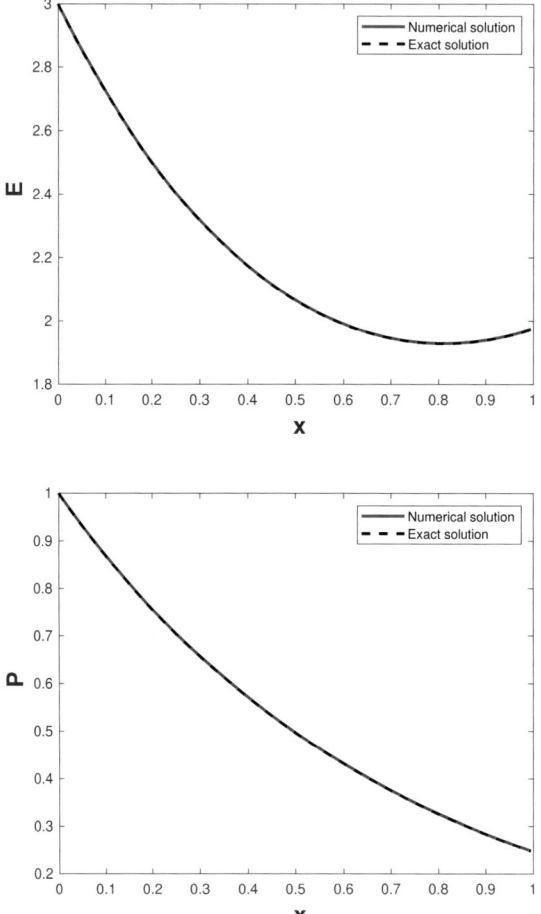

Figure 2.10: 1D moving equilibrium: profile of the energy (top), and pressure (bottom) obtained at time $t = 10$.

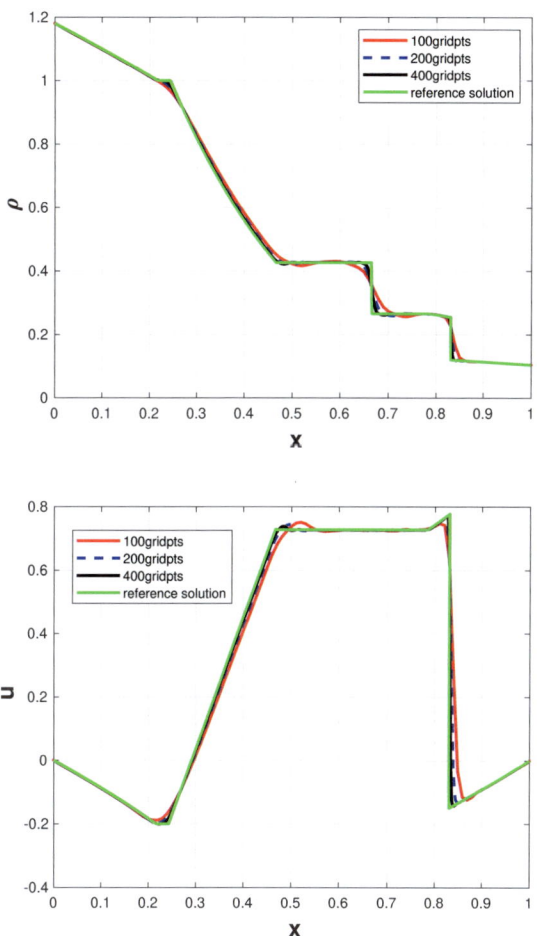

Figure 2.11: 1D shock tube problem: density (top), velocity (bottom) at time $t = 0.2$.

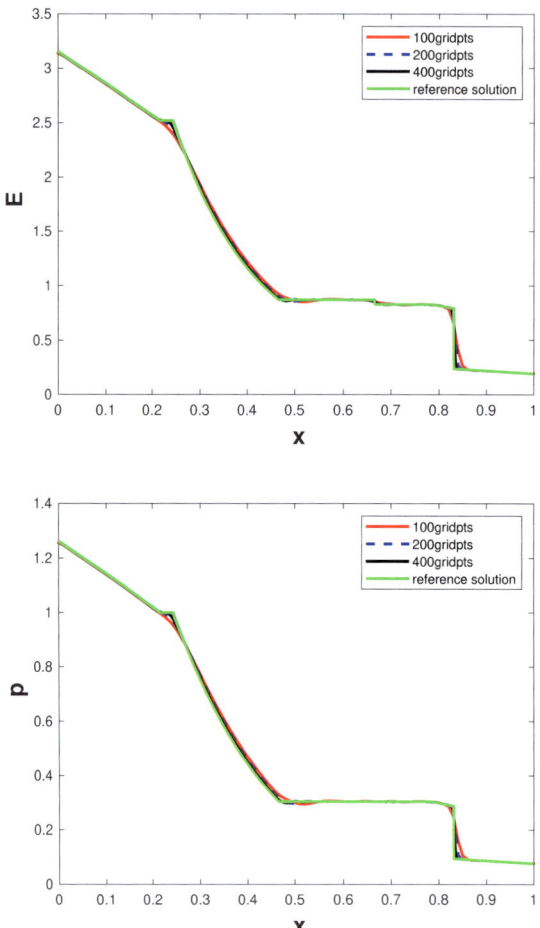

Figure 2.12: 1D shock tube problem: energy (top), pressure (bottom) at time $t = 0.2$.

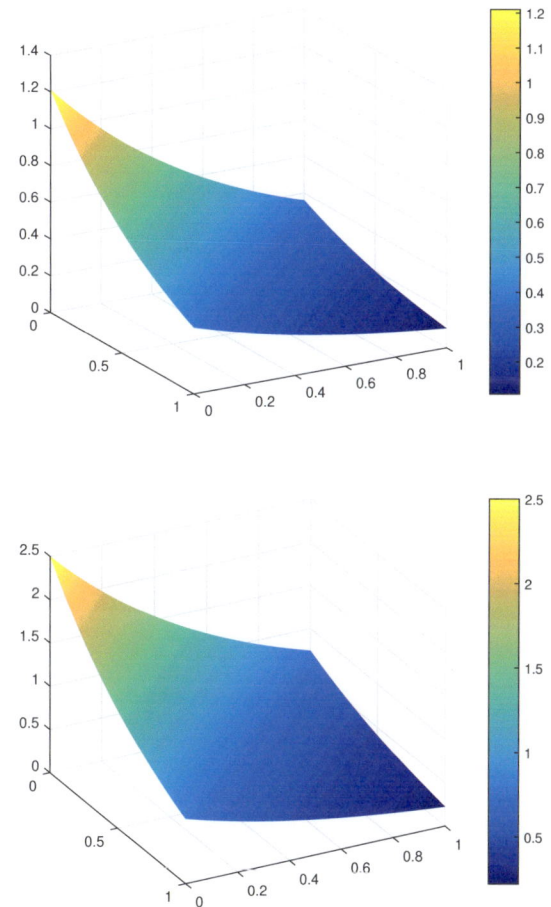

Figure 2.13: 2D isothermal equilibrium: density (left), energy (right) obtained at the final time $t = 0.25$.

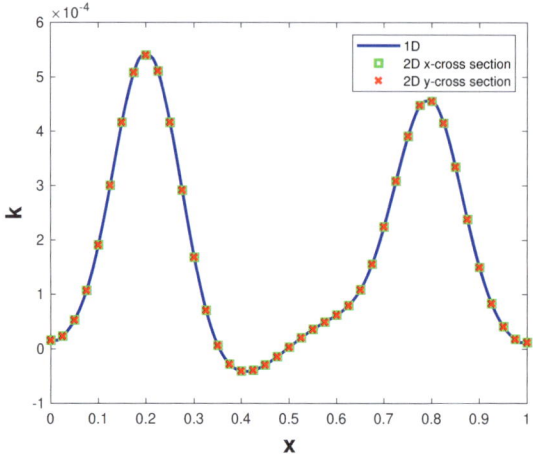

Figure 2.14: 2D Unidirectional equilibrium perturbation: 1D/2D comparison of the pressure perturbation k at time $t = 0.25$.

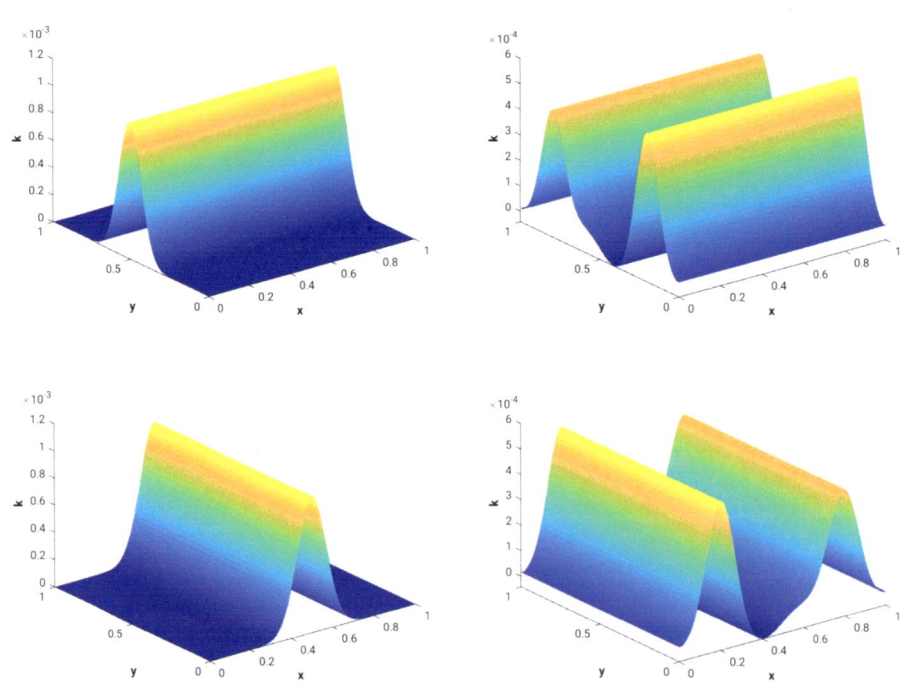

Figure 2.15: 2D Unidirectional equilibrium perturbation: initially along x (top left), at $t = 0.25$ along x (top right), initially along y (bottom left), at $t = 0.25$ along y (bottom right).

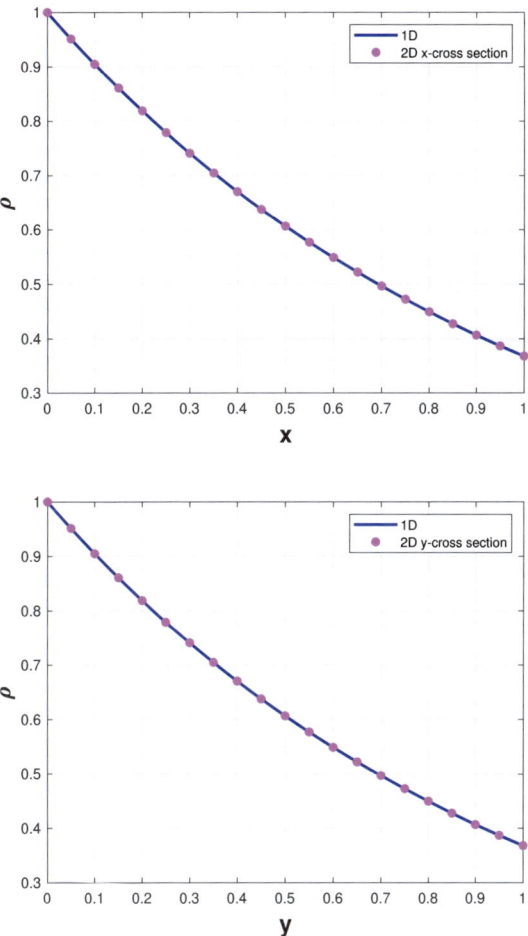

Figure 2.16: 2D moving equilibrium: the density at time $t = 0.25$ with 1D /2D x-cross section (left) and 1D /2D y-cross section (right).

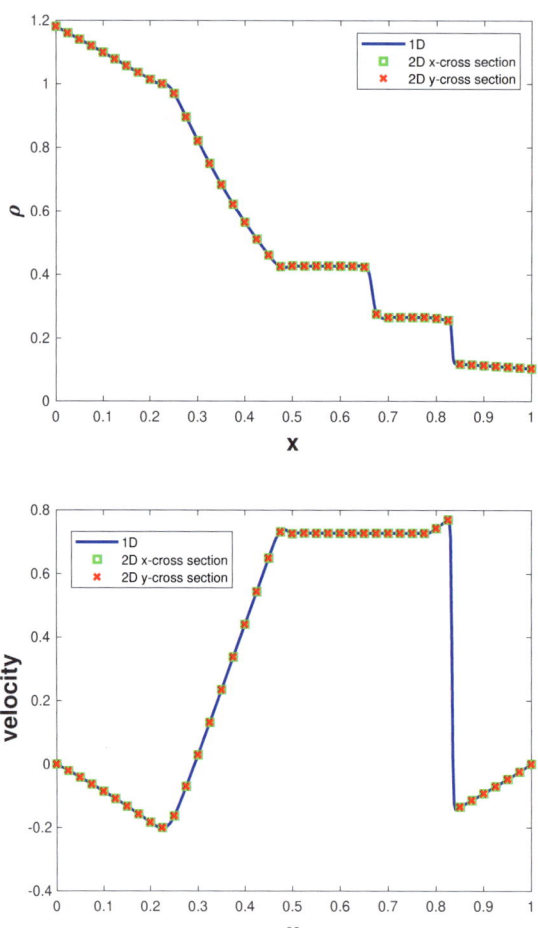

Figure 2.17: 2D shock tube problem: 1D-2D comparison density (left) and velocity (right) at time $t = 0.2$.

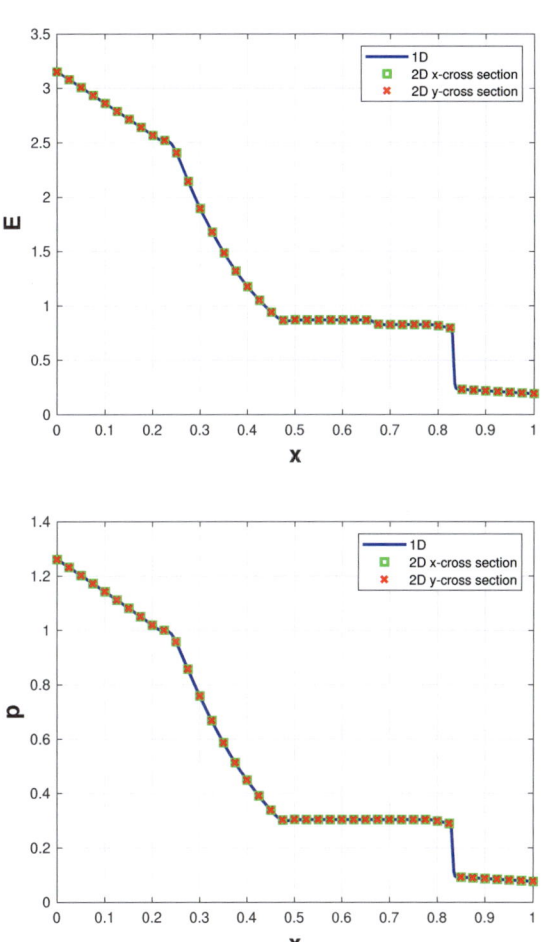

Figure 2.18: 2D shock tube problem: 1D-2D comparison energy (left), pressure (right) at time $t = 0.2$.

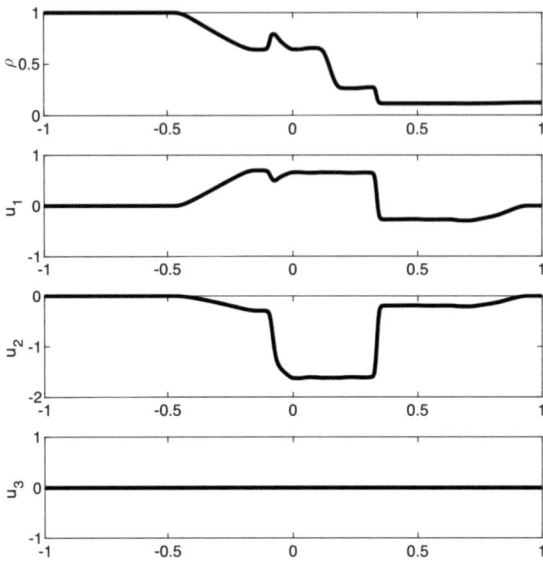

Figure 2.19: 2D shock tube problem: cross sections of the density and the velocity components at time $t = 0.25$.

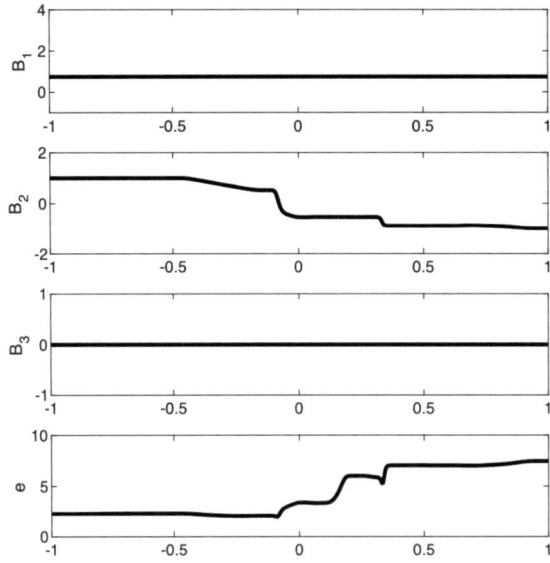

Figure 2.20: 2D shock tube problem: cross sections of the energy and the magnetic field components at time $t = 0.25$.

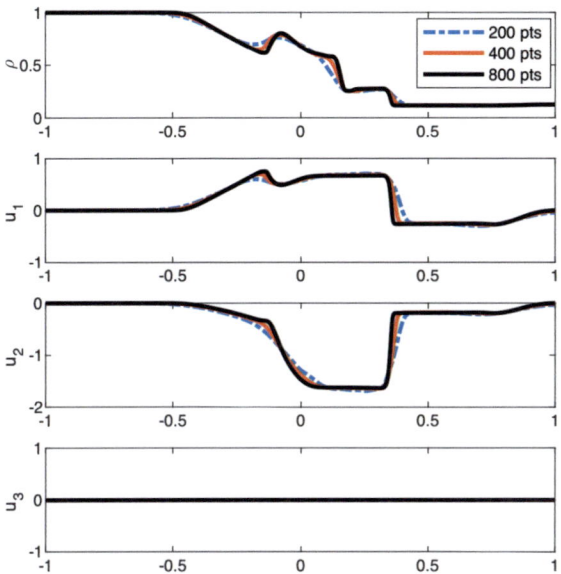

Figure 2.21: 2D shock tube problem: cross sections of the density and the velocity components at time $t = 0.25$ on 200×200 (dashed line), 400×400 (solid red line), and 800×800 (solid black line) grid points.

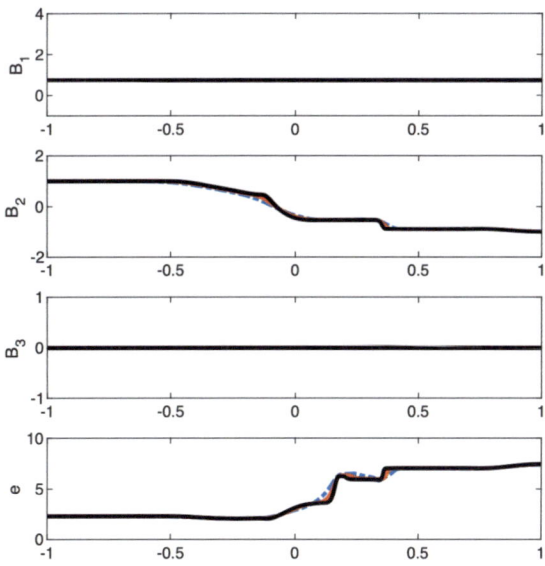

Figure 2.22: 2D shock tube problem: cross sections of the energy and the magnetic field components at time $t = 0.25$ on 200×200 (dashed line), 400×400 (solid red line), and 800×800 (solid black line) grid points.

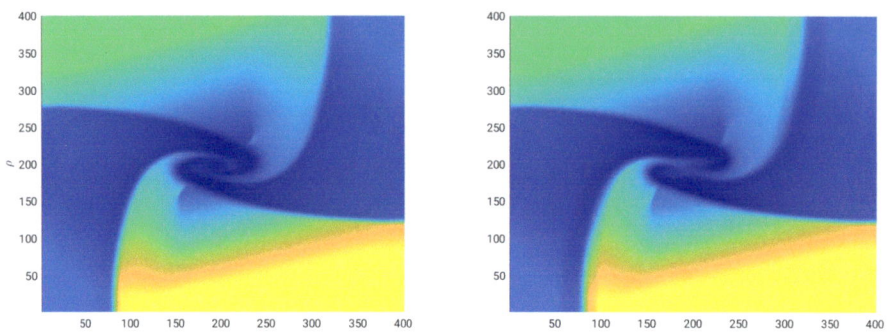

Figure 2.23: Four stages Riemann problem: ρ with CTM (left) and without CTM (right) at the final time $t = 0.8$.

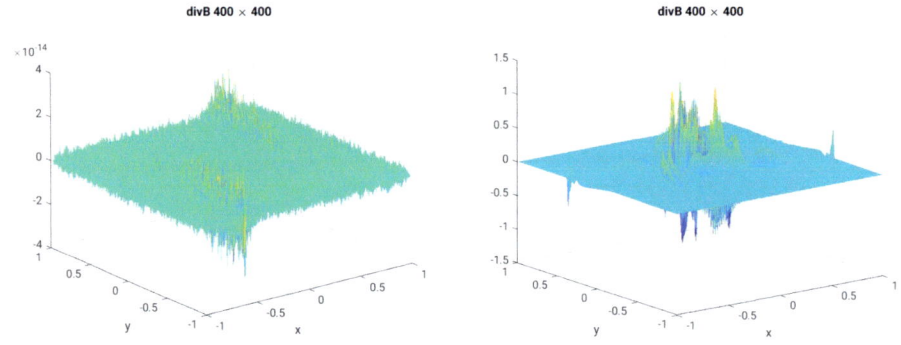

Figure 2.24: Four stages Riemann Problem: div**B** with CTM (left) and without CTM (right) at the final time $t = 0.8$.

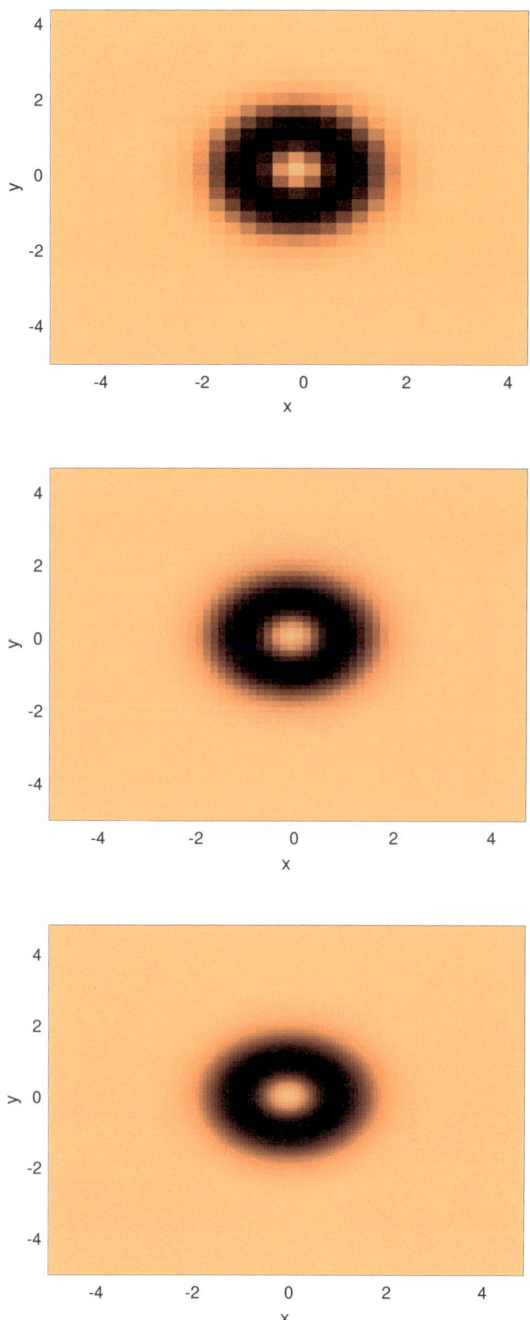

Figure 2.25: MHD vortex: pressure profile at the final time on 32×32, 64×64 and 128×128 grid points respectively.

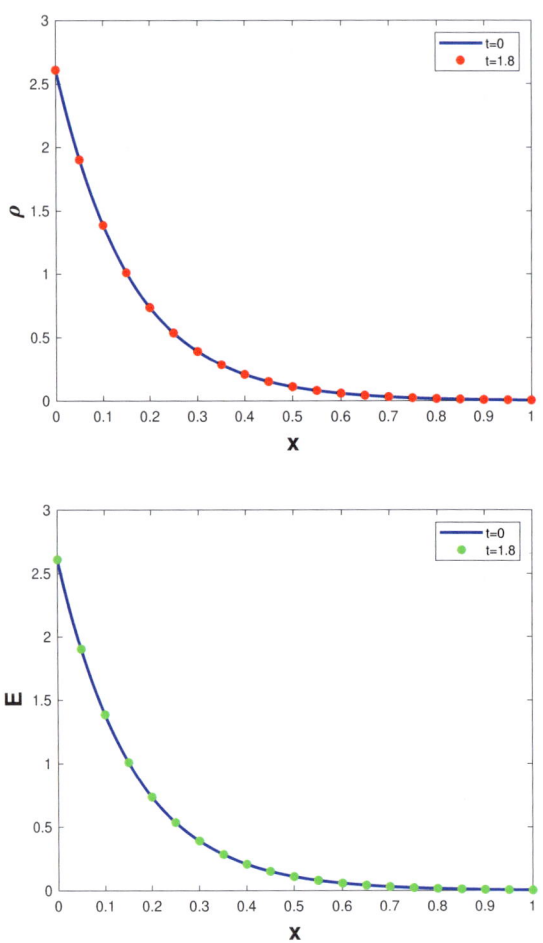

Figure 2.26: Hydrodynamic wave propagation: a comparison of the cross sections of the density ρ (left) and the energy E (right) initially and at the final time $t = 1.8$.

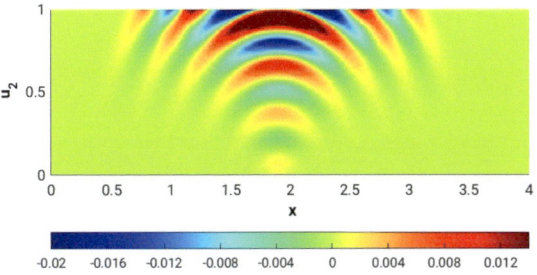

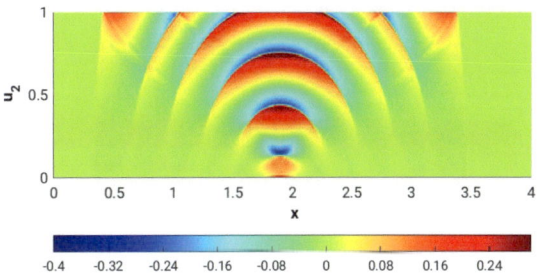

Figure 2.27: Hydrodynamic wave propagation: wave profile u_2 for $c = 0.003$ (left) and $c = 0.3$ (right) at the final time $t = 1.8$.

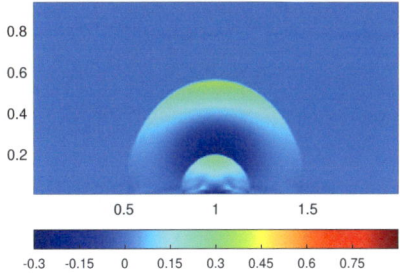

Figure 2.28: MHDwave propagation: velocity in a direction parallel to the magnetic field $u_B = <\mathbf{u}, \mathbf{B}> / |\mathbf{B}|$ for $\mu = 0$ on 400×200 grid points at the final time $t = 0.54$.

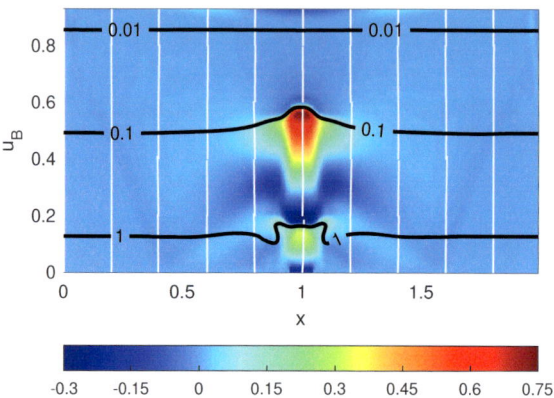

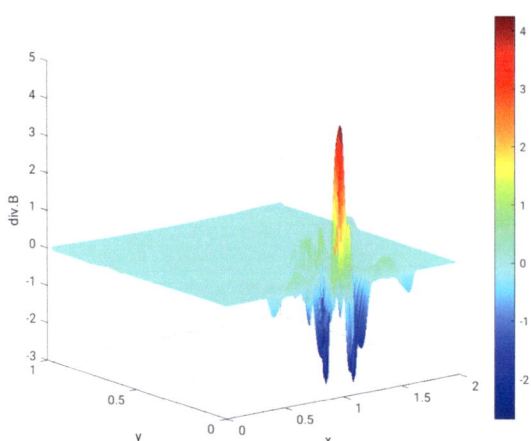

Figure 2.29: MHDwave propagation: velocity in a direction parallel to the magnetic field $u_B = <\mathbf{u}, \mathbf{B}> /|\mathbf{B}|$ for $\mu = 1$ on 400×200 grid points at the final time $t = 0.54$ without CTM.

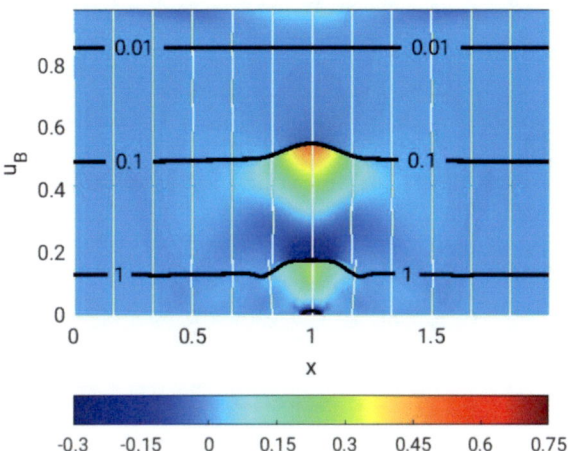

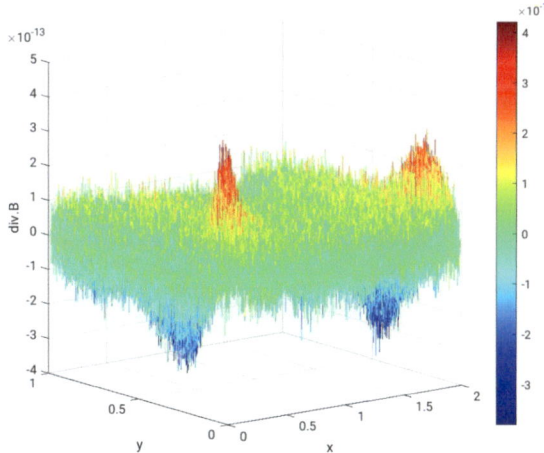

Figure 2.30: MHD wave propagation: velocity in a direction parallel to the magnetic field $u_B =< \mathbf{u}, \mathbf{B} > /|\mathbf{B}|$ for $\mu = 1$ on 1200×600 grid points at the final time $t = 0.54$ with CTM.

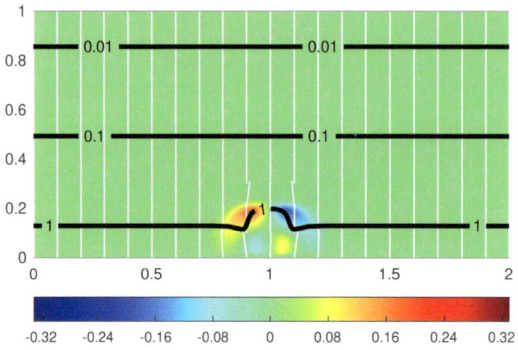

(a) $u_{\perp B}$ at t $= 0.216$

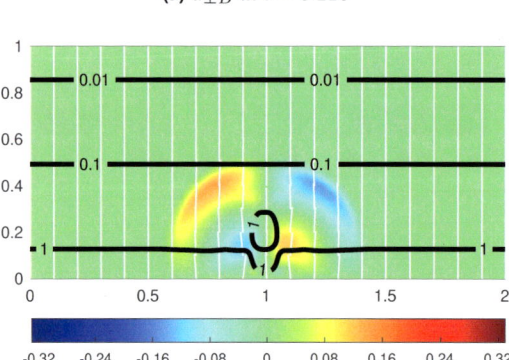

(b) $u_{\perp B}$ at t $= 0.36$

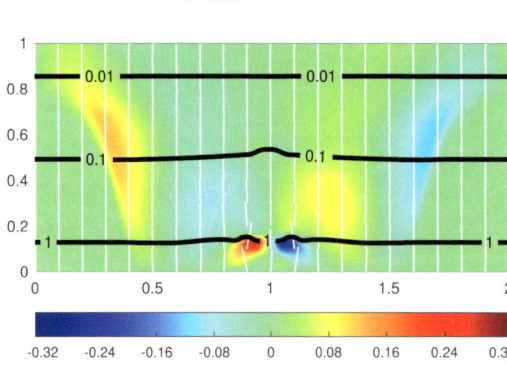

(c) $u_{\perp B}$ at t $= 0.504$

Figure 2.31: MHD wave propagation: $u_{\perp B} = < (u_1, u_2), (-B_2, B_1) > /|\mathbf{B}|$ for $\mu = 1$ on 400×200 grid points at different times.

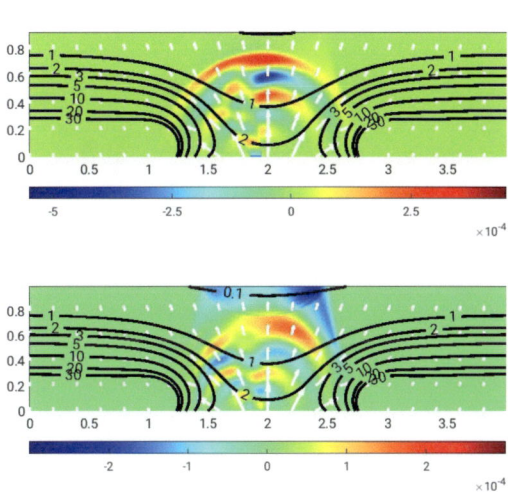

Figure 2.32: MHD wave propagation - weak magnetic field: velocity in a direction parallel to the magnetic field $u_B =< \mathbf{u}, \mathbf{B} > /|\mathbf{B}|$ (left) and perpendicular to the magnetic field $u_{\perp B} =< (u_1, u_2), (-B_2, B_1) > /|\mathbf{B}|$ (right) on 800×200 grid points at the final time $t = 0.9$.

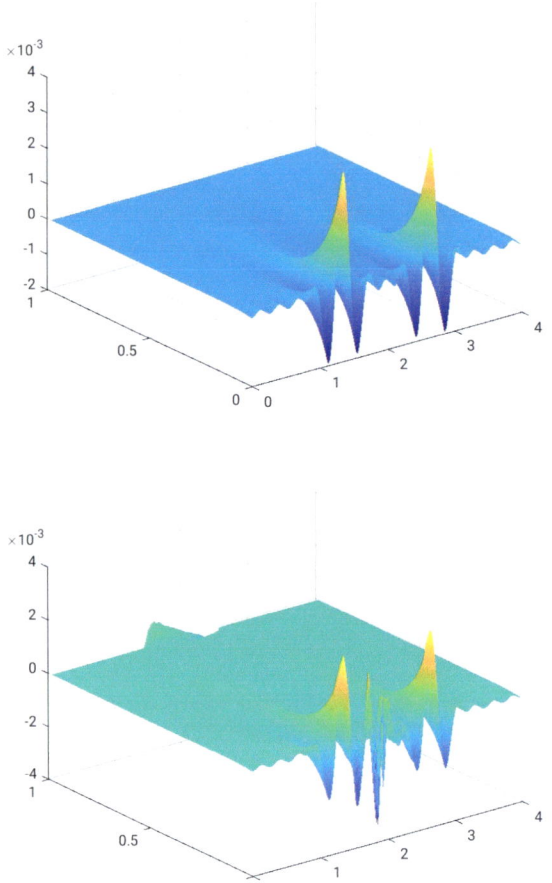

Figure 2.33: div·**B** initially at $t = 0$ (left) and at the final time $t = 0.9$ (right).

Chapter 3

AP and SP Schemes for Kinetic equations

3.1   Introduction

After working on well-balancing techniques for the Euler equations. We started looking at possibilities to generalize the well-balancing approach. Digging deeper into the derivation of the Euler equations, we noticed a connection between the Euler equations and the kinetic models such that, rescaled kinetic models (as we will see in the following sections) converge to fluid equations. When the average distance between two successive velocity changes is small, i.e., the mean free path is small, one has to use resolved space and time steps that are less than the mean free path. Moreover, the probability density function in kinetic models depends not only on space and time, but also on velocity. The high dimensionality and the small mean free path led to an extremely high computational cost, and AP schemes that allow mean free path independent meshes became popular in the last decades.

AP schemes were first proposed in [57, 26] for the neutron transport equation and have been successfully extended to a lot of applications, we refer to the review paper [47] for more discussions. Different AP schemes have been developed for various kinetic models, including the neutron transport equation [1, 49, 57, 60], the velocity jump model for E.Coli chemotaxis [15, 21], and the Boltzmann equation [29, 54, 16, 41]. The Knudsen number ε is the ratio of the mean free path and the domain typical length scale [26]. To prove that a scheme is AP, one has to show that when the Knudsen number goes to zero in the discretized scheme, it converges to a good discretization of the corresponding limit model. The main advantage of AP schemes is that their stability and convergence are independent of the Knudsen number. For such models, since the equilibrium is not known at the begining, it can only be reached after a certain amount of time, which means it is not known and cannot be initially given as the well-balanced techniques require. Moreover, as the parameter in the equation takes a new value, a new equilibrium pops up. Hence, the common well-balanced techniques will not be useful here and the need for stationary preserving schemes, as mentioned before, arises. The investigation first adressed two questions: how can we see the SP property for the corresponding AP schemes; and how can we project what we understand at the kinetic level back to the fluid level. Our key observation is that, as far as the Maxwellian of the distribution function of an AP scheme can be updated explicitly, the second requirement of the SP property is satisfied immediately. Our proof of the SP property is independent of ε and applicable whenever the discretization linearly depends on the Maxwellian of the collision operator. Numerically, one can check that the time evolutionary problem converges to a discrete stationary solution after finite time, and their difference is smaller than machine precision. In the subsequent part, we will consider three different classes of AP schemes for which one can prove their SP properties as examples on which our criterion applies. Once we are able to show that for an AP scheme, the space and velocity discretization of the stationary equation provides a good approximation to the steady state solution for all ε, and the Maxwellian of the distribution function is updated explicitly, the SP property follows immediately. To show the universality of our observation, we test different kinetic models for different AP schemes, as listed in Table 3.1.

For this aim, we considered three schemes for three different kinetic models in sections 3.2, 3.3 and 3.4; see table 3.1. We tried to prove the SP property for each scheme separately and a useful conclusion has been drawn [27]. In each section, we present the kinetic model and its corresponding AP scheme with the SP property followed by some numerical evidence. For the velocity space in the numerical test cases, the standard Gaussian quadrature set is used.

Kinetic Model	Scheme
Neutron transport equation	Parity-equations based
Chemotaxis kinetic model	UGKS
Boltzmann equation	IMEX Penalization method

Table 3.1: A list of kinetic models together with their corresponding schemes.

3.2 Parity equations-based scheme for the Neutron transport equation

In this section we check the parity equations-based AP scheme for the neutron transport equation in [47, 48]. This scheme is then proven to be SP as well.

3.2.1 The neutron transport equation

Consider the 1D neutron transport equation:

$$\partial_t f + \frac{1}{\varepsilon} v \cdot \nabla_x f = \frac{\sigma_T}{\varepsilon^2} \left(\frac{1}{2} \int_{-1}^{1} f dv' - f \right) - \sigma_a \left(\frac{1}{2} \int_{-1}^{1} f dv' \right) + q \tag{3.1}$$

with $x \in [x_L, x_R]$ and $v \in [-1, 1]$. $f = f(t, x, v)$ is the particle distribution function and v is the particle velocity. We present the scheme for a simplified neutron transport equation with $\sigma_T = 1, \sigma_a = 0, q = 0$. The extension to more general cases does not add any difficulties.

3.2.2 Discretization of the model

When $\sigma_T = 1, \sigma_a = 0, q = 0$ in (3.1), the parity equations-based scheme in [48] can be summarized by the following steps:

- Rewrite (3.1) into two equations. For $v \geq 0$,

$$\varepsilon \partial_t f(v) + v \partial_x f(v) = \frac{1}{\varepsilon}\left(\frac{1}{2}\int_{-1}^1 f \, dv - f(v)\right),$$

$$\varepsilon \partial_t f(-v) - v \partial_x f(-v) = \frac{1}{\varepsilon}\left(\frac{1}{2}\int_{-1}^1 f \, dv - f(-v)\right). \tag{3.2}$$

- Introduce the even and odd parities that are

$$r(t,x,v) = \frac{1}{2}[f(t,x,v) + f(t,x,-v)], \qquad j(t,x,v) = \frac{1}{2\varepsilon}[f(t,x,v) - f(t,x,-v)].$$

- Add and subtract the equations in (3.2) and rewrite them into the following diffusive relaxation system,

$$\partial_t r + v \partial_x j = -\frac{1}{\varepsilon^2}(r - \rho_r),$$

$$\partial_t j + \eta v \partial_x r = -\frac{1}{\varepsilon^2}[j + (1 - \epsilon^2\eta)v\partial_x r], \tag{3.3}$$

where $\rho_r = \int_0^1 r \, dv'$ and $\eta = \eta(\varepsilon)$ is such that, $0 \leq \eta \leq \frac{1}{\varepsilon^2}$ in order to guarantee the positivity of $\eta(\varepsilon)$ and $(1 - \epsilon^2\eta(\varepsilon))$ so the problem remains well-posed uniformly in ε. η is chosen as $\eta(\varepsilon) = \min(1, \frac{1}{\varepsilon})$.

- Split the equations (3.3) into two steps:

  - Relaxation step:

  $$\begin{cases} \partial_t r = -\frac{1}{\epsilon^2}(r - \rho_r), \\ \partial_t j = -\frac{1}{\epsilon^2}[j + (1 - \epsilon^2\eta)v\partial_x r]. \end{cases}$$

  - Transport step:

  $$\begin{cases} \partial_t r + v \partial_x j = 0, \\ \partial_t j + \eta v \partial_x r = 0. \end{cases}$$

- Discretize the two steps as follows:

  - For the transport step, we use an explicit first order upwind scheme on its diagonal form such that

  $$\begin{cases} r_i^{n+\frac{1}{2}} = r_i^n - v\frac{\Delta t}{\Delta x}D^u j_i^n, \\ j_i^{n+\frac{1}{2}} = j_i^n - \eta v\frac{\Delta t}{\Delta x}D^u r_i^n, \end{cases} \tag{3.4}$$

  where $D^u f_i^n = f_{i+1}^n - f_i^n$ and $D^c f_i^n = \frac{f_{i+1}^n - f_{i-1}^n}{2}$ are the upwind and the central spatial differences respectively.

– For the relaxation step, we use an implicit backward Euler method that writes

$$\begin{cases} \frac{r_i^{n+1}-r_i^{n+\frac{1}{2}}}{\Delta t} = -\frac{1}{\varepsilon^2}\left(r_i^{n+1} - \rho_{r_i}^{n+1}\right), \\ \frac{j_i^{n+1}-j_i^{n+\frac{1}{2}}}{\Delta t} = -\frac{1}{\varepsilon^2}\left(j_i^{n+1} + (1 - \epsilon^2\eta)v\frac{D^c}{\Delta x}r_i^{n+1}\right). \end{cases}$$

By integrating the above first equation over V (the velocity space), we find, $\rho_{r_i}^{n+1} = \rho_{r_i}^{n+\frac{1}{2}}$. Then,

$$\begin{cases} r_i^{n+1} = Ar_i^{n+\frac{1}{2}} + B\rho_{r_i}^{n+\frac{1}{2}}, \\ j_i^{n+1} = Aj_i^{n+\frac{1}{2}} - B(1 - \varepsilon^2\eta)v\frac{D^c}{\Delta x}r_i^{n+1}, \end{cases} \tag{3.5}$$

with A and B being defined as:

$$A = \frac{\varepsilon^2}{\varepsilon^2+\Delta t} \quad \text{and} \quad B = \frac{\Delta t}{\varepsilon^2+\Delta t}.$$

The fully space-time discretized parity equations-based AP scheme is given by the transport step (3.4) and the relaxation step (3.5). The boundary conditions for r and j are the same as in [48] and are obtained using the following relations:

$$r + \varepsilon j|_{x=x_L} = F_L(v) \ and \ r - \varepsilon j|_{x=x_R} = F_R(v) \tag{3.6}$$

when $\varepsilon << 1$, j can be approximated by,

$$j = -v\partial_x r \tag{3.7}$$

from the second equation in (3.3). Hence, the boundary conditions for r and j are,

$$r - \varepsilon v\partial_x r|_{x=x_L} = F_L(v) \ and \ r + \varepsilon v\partial_x r|_{x=x_R} = F_R(v) \tag{3.8}$$

$$j = -v\partial_x r \tag{3.9}$$

where $F_L(v)$ and $F_R(v)$ are the inflow boundary conditions of f. The AP proof of the scheme has previously been done [48], [47], [15].

3.2.3 SP property

The purpose of this section is to prove that the scheme has the SP property. As mentioned in the introduction, the scheme has to meet two requirements. The first requirement is satisfied when an AP discretization of the steady state equation is provided. The proof is given in Appendix B. For the second requirement, we need to prove that starting from a discrete stationary solution, the solution of the time evolutionary problem does not change. Plugging (3.4) in (3.5) and using the fact that $\rho_r^{n+\frac{1}{2}} = \rho_r^{n+1}$, the equations for updating r^{n+1} and j_i^{n+1} can be written as:

$$\frac{r_i^{n+1} - r_i^n}{\Delta t} + v\frac{D^u}{\Delta x}j_i^n = -\frac{1}{\varepsilon^2}(r_i^{n+1} - \rho_{r_i}^{n+1}), \tag{3.10a}$$

$$\frac{j_i^{n+1} - j_i^n}{\Delta t} + \eta v\frac{D^u}{\Delta x}r_i^n = -\frac{1}{\varepsilon^2}(j_i^{n+1} + (1 - \varepsilon^2\eta)v\frac{D^c}{\Delta x}r_i^{n+1}). \tag{3.10b}$$

A discrete stationary solution to (3.10) are r_i^n and j_i^n that satisfies:

$$v\frac{D^u}{\Delta x}j_i^n = -\frac{1}{\varepsilon^2}(r_i^n - \rho_{r_i}^n), \tag{3.11a}$$

$$\eta v\frac{D^u}{\Delta x}r_i^n = -\frac{1}{\varepsilon^2}[j_i^n + (1 - \varepsilon^2\eta)v\frac{D^c}{\Delta x}r_i^n]. \tag{3.11b}$$

Lemma 1. *When r_i^n and j_i^n are solutions of the steady state equation discretization (3.11), then $r_i^{n+1} = r_i^n$ and $j_i^{n+1} = j_i^n$. Hence the parity equations-based scheme satisfies the second requirement of the SP property.*

Proof.

- For r: Since $\rho_{r_i}^n = \int_0^1 r_i^n$, inserting (3.11a) in (3.10a) and integrating over $[0, 1]$ yields $\rho_r^{u+1} = \rho_r^n$.
  Using (3.11a) and $\rho_r^{n+1} = \rho_r^n$, (3.10a) gives

$$\frac{r_i^{n+1} - r_i^n}{\Delta t} - \frac{1}{\varepsilon^2}(r_i^n - \rho_{r_i}^n) = -\frac{1}{\varepsilon^2}(r_i^{n+1} - \rho_{r_i}^n).$$

Hence,

$$(\frac{1}{\Delta t} + \frac{1}{\epsilon^2})(r_i^{n+1} - r_i^n) = 0.$$

and then $r_i^{n+1} = r_i^n$.

- For j: Using $r^{n+1} = r^n$, (3.10b) becomes

$$\frac{j_i^{n+1} - j_i^n}{\Delta t} + \eta v\frac{D^u}{\Delta x}r_i^n = -\frac{1}{\epsilon^2}[j_i^{n+1} + (1 - \epsilon^2\eta)v\frac{D^c}{\Delta x}r_i^n]. \tag{3.12}$$

From (3.11b), (3.12) writes,

$$\frac{j_i^{n+1} - j_i^n}{\Delta t} - \frac{1}{\epsilon^2}[j_i^n + (1 - \epsilon^2\eta)v\frac{D^c}{\Delta x}r_i^n] = -\frac{1}{\epsilon^2}[j_i^{n+1} + (1 - \epsilon^2\eta)v\frac{D^c}{\Delta x}r_i^n].$$

Then,

$$(\frac{1}{\Delta t} + \frac{1}{\epsilon^2})(j_i^{n+1} - j_i^n) = 0$$

and thus $j_i^{n+1} = j_i^n$.

$\square$

Using this Lemma, the scheme satisfies both requirements of the SP property as mentioned in the bullet points in the introduction. This is because, starting from a discrete stationary solution, our discretization of the time evolutionary problem does not change this discrete stationary solution. This way we have shown that the parity equations-based scheme (which is AP) has both the AP and SP properties.

3.2.4 Numerical results

To validate the AP and SP properties of the parity equations-based scheme, we use the same initial and boundary conditions as problem 1 in section 6 in [48]. The initial condition, given by the distribution function is $f(x, v, t = 0) = 0$, and the computational domain is $x \in [0, 1]$. The boundary conditions are as in (3.8) and (3.9) with

$$F_L(v) = 1 \quad \text{and} \quad F_R(v) = 0.$$

This data are consistent as can be seen by (3.8) and (3.9). The mesh and time step sizes are respectively $\Delta x = 0.025$ and $\Delta t = 0.0002$ with the S_{16} Gaussian quadrature points for the velocity. In figure 3.1, we plot the density at time $t = 0.05$ for $\varepsilon = 10^{-2}$, $\varepsilon = 10^{-3}$, $\varepsilon = 10^{-6}$ and compare it to its diffusion limit. The curves get close to each other when ε gets very small. The curve corresponding to $\varepsilon = 10^{-6}$ is exactly on top of the curve of the diffusion limit equation. This verifies the AP property of the scheme. Furthermore, we plot in figure 3.2 the time evolution of the distance between the numerical stationary solution ρ_r^s and the numerical solution ρ_r of the time evolutionary equation given by the L^∞ norm

$$||\rho_r - \rho_r^s||_\infty = \max_j \{\rho_{rj} - \rho_{rj}^s\}.$$

One can see that this distance does not change after we reach the steady state. After that we give the norm at discrete times in Table 3.2 where we also show that the SP property is valid for all $\varepsilon \ll 1$. Figure 3.2 and Table 3.2 indicate that the SP property is well satisfied.

T	0	2	4	6	8
L^∞	0.995	1.051×10^{-3}	1.683×10^{-6}	2.696×10^{-9}	4.120×10^{-12}
T	0	2	4	6	8
L^∞	1	9.111×10^{-4}	1.263×10^{-6}	1.752×10^{-9}	2.176×10^{-12}

Table 3.2: Neutron Transport: L^∞-norm of the difference between the solution and the stationary solution in the time interval [0,8] for $\varepsilon = 10^{-2}$ (top) and $\varepsilon = 10^{-8}$ (bottom).

3.3 UGKS scheme for the chemotaxis kinetic model

In this section we first extend the UGKS in [60, 54, 55] to the time evolutionary chemotaxis model, then show its AP and SP properties. The AP scheme is derived by Min Tang and Casimir Emako.

3.3.1 The chemotaxis kinetic model

The chemotaxis kinetic model models bacteria that undergo run and tumble process as mentioned in [40, 70, 71]. During the run phase, bacteria move along a straight line and change their directions during the tumble phase.

This is called the velocity jump process and can be modeled by the Othmer-Dunbar-Alt model that writes [2, 62]:

$$\begin{cases} \partial_t f + \frac{1}{\varepsilon} v \cdot \nabla_x f = \frac{1}{\varepsilon^2}[\frac{1}{|V|}\int_V (1 + \varepsilon\phi(v' \cdot \partial_x S))f(v')dv' - (1 + \varepsilon\phi(v \cdot \partial_x S))f(v)], \\ \partial_t S - D\Delta S + \alpha S = \beta\rho, \quad \rho(x,t) := \frac{1}{|V|}\int_V f(v)dv. \end{cases}$$
(3.13)

Here $f(x,v,t)$ is the probability density function at time t, position x and moving with velocity v; ϕ is an odd decreasing function such that $\phi(-u) = -\phi(u)$; $S(x,t)$ is the concentration of a chemical substance where the parameters D, α, β are positive constants; ε is the Knudsen number. When $\phi = 0$, the chemotaxis kinetic model reduces to the neutron transport equation. As $\varepsilon \to 0$, $f(x,v,t)$ converges to $\rho_0(x,t)$, where $\rho_0(x,t)$ solves the following Keller-Segel equation [17, 42, 63]:

$$\begin{cases} \partial_t \rho_0 = \frac{1}{3}\Delta\rho_0 + \nabla((\frac{1}{|V|}\int_V v\phi(v\partial_x S)dv)\rho_0), \\ \partial_t S - D\Delta S + \alpha S = \beta\rho_0. \end{cases}$$
(3.14)

3.3.2 Discretization of the model

Before discussing about the more complex equation for f, we first discretize the equation for the chemical concentration S. Let $S_i^n \approx S(x_i, t^n)$, the following centered finite difference method is used to update S:

$$\frac{S_i^{n+1} - S_i^n}{\Delta t} = D\frac{S_{i+1}^{n+1} - 2S_i^{n+1} + S_{i-1}^{n+1}}{\Delta x^2} - \alpha S_i^{n+1} + \beta\rho_i^n.$$
(3.15)

After S_i^{n+1} is obtained, we approximate $\partial_x S^{n+1}$ by a piecewise constant function such that

$$\partial_x S(x, t^{n+1}) \approx \partial_x S(x_{i+\frac{1}{2}}, t^{n+1}) \approx \frac{S_{i+1}^{n+1} - S_i^{n+1}}{\Delta x} := \sigma_{i+\frac{1}{2}}, \qquad \text{for } \forall x \in [x_i, x_{i+1}).$$
(3.16)

The UGKS is a finite volume approach for discretizing the kinetic equation of f. By integrating the chemotaxis kinetic model (3.13) over $[x_{i-\frac{1}{2}}, x_{i+\frac{1}{2}}] \times [t^n, t^{n+1}] \times V$ and letting $f_i^n = \frac{1}{\Delta x}\int_{x_{i-\frac{1}{2}}}^{x_{i+\frac{1}{2}}} f(x,v,t^n)\,dx$, $\rho_i^n = \frac{1}{|V|}\int_V f_i^n\,dv$, the total density ρ_i^{n+1}

67

and density fluxes f_i^{n+1} is updated as follows

$$\frac{\rho_i^{n+1} - \rho_i^n}{\Delta t} + \frac{F_{i+\frac{1}{2}}^n - F_{i-\frac{1}{2}}^n}{\Delta x} = 0, \tag{3.17}$$

$$\frac{f_i^{n+1} - f_i^n}{\Delta t} + \frac{\Phi_{i+\frac{1}{2}}^n - \Phi_{i-\frac{1}{2}}^n}{\Delta x} = \frac{1}{\varepsilon^2}\left(\rho_i^{n+1} - f_i^{n+1}\right)$$

$$+ \frac{1}{\varepsilon}\left(\frac{1}{|V|}\int_V \phi(v'\sigma_{i+\frac{1}{2}})f_i^n(v')\,dv' - \phi(v\sigma_{i+\frac{1}{2}})f_i^n\right). \tag{3.18}$$

Here the numerical fluxes are given by

$$\Phi_{i+\frac{1}{2}}^n = \frac{1}{\varepsilon\Delta t}\int_{t^n}^{t^{n+1}} vf(x_{i+\frac{1}{2}}, v, t)\,dt,$$

$$F_{i+\frac{1}{2}}^n = \frac{1}{|V|}\int_V \left(\frac{1}{\varepsilon\Delta t}\int_{t^n}^{t^{n+1}} vf(x_{i+\frac{1}{2}}, v, t)\,dt\right)dv. \tag{3.19}$$

It is important to note that $\sigma_{i+\frac{1}{2}}$ approximates $\partial_x S$ in the interval $[x_i, x_{i+1})$, while f_i^n is the average density over the cell $[x_{i-\frac{1}{2}}, x_{i+\frac{1}{2}})$. This choice is important to obtain the correct advection term in the limit Keller-Segel model when ε becomes small. We use discrete ordinate method for the velocity discretization, but for simplicity, we write the scheme in continuous velocity. The most crucial step for UGKS is to determine $\Phi_{i+\frac{1}{2}}^n$ and $F_{i+\frac{1}{2}}^n$. The details are listed below:

- **Find the approximation of** $f(x_{i+\frac{1}{2}}, v, t)$. The 1D chemotaxis model (3.13) can be rewritten as:

$$\partial_t f + \frac{1 + \varepsilon\phi(v\partial_x S^\varepsilon)}{\varepsilon^2}f + \frac{v}{\varepsilon}\partial_x f = \frac{1}{\varepsilon^2}\mathcal{T}^1 f, \tag{3.20}$$

where $(\mathcal{T}^1 f)(x,t) := \frac{1}{|V|}\int_V \left(1 + \varepsilon\phi(v'\partial_x S)\right)f(x, v', t)dv'$. Consider the interval $[x_i, x_{i+1})$, multiplying both sides of (3.20) by $\exp\left(\frac{(1+\varepsilon\phi(v\sigma_{i+\frac{1}{2}})}{\varepsilon^2}t\right)$ yields

$$\frac{d}{dt}\left[f(x + \frac{v}{\varepsilon}t, v, t)\exp\left(\frac{(1 + \varepsilon\phi(v\sigma_{i+\frac{1}{2}})}{\varepsilon^2}t\right)\right]$$

$$= \frac{\mathcal{T}^1 f(x,t)}{\varepsilon^2}\exp\left(\frac{(1 + \varepsilon\phi(v\sigma_{i+\frac{1}{2}})}{\varepsilon^2}t\right).$$

Integrating the above equation over (t^n, t) yields to,

$$f(x_{i+\frac{1}{2}}, v, t) = f(x_{i+\frac{1}{2}} - \frac{v}{\varepsilon}(t - t^n), v, t^n)\exp\left(-\frac{(1 + \varepsilon\phi(v\sigma_{i+\frac{1}{2}})}{\varepsilon^2}(t - t^n)\right)$$

$$+ \frac{1}{\varepsilon^2}\int_{t^n}^t \mathcal{T}^1 f(x_{i+\frac{1}{2}} - \frac{v}{\varepsilon}(t - s), s)\exp\left(-\frac{(1 + \varepsilon\phi(v\sigma_{i+\frac{1}{2}})}{\varepsilon^2}(t - s)\right)ds. \tag{3.21}$$

This is an exact expression for $f(x_{i+\frac{1}{2}}, v, t)$ that will be used to determine $\Phi^n_{i+\frac{1}{2}}$, $F^n_{i+\frac{1}{2}}$ in (3.19). At this stage, we need to approximate $f(x, v, t^n)$ and $(\mathcal{T}^1 f)(x, t)$ on the right hand side of (3.21). f is approximated by a piecewise constant function and $\mathcal{T}^1 f$ by a piecewise linear function as follows:

$$f(x, v, t^n) = \begin{cases} f_i^n, & x < x_{i+\frac{1}{2}}, \\ f_{i+1}^n, & x > x_{i+\frac{1}{2}}, \end{cases}$$

$$\mathcal{T}^1 f(x, t) = \begin{cases} \mathcal{T}^1 f_{i+\frac{1}{2}}^n + \delta^L \mathcal{T}^1 f_{i+\frac{1}{2}}^n (x - x_{i+\frac{1}{2}}), & x < x_{i+\frac{1}{2}}, \\ \mathcal{T}^1 f_{i+\frac{1}{2}}^n + \delta^R \mathcal{T}^1 f_{i+\frac{1}{2}}^n (x - x_{i+\frac{1}{2}}), & x > x_{i+\frac{1}{2}}. \end{cases}$$

Here, $\mathcal{T}^1 f_{i+\frac{1}{2}}^n$, $\delta^L \mathcal{T}^1 f_{i+\frac{1}{2}}^n$, and $\delta^R \mathcal{T}^1 f_{i+\frac{1}{2}}^n$ are defined by:

$$\begin{cases} \mathcal{T}^1 f_{i+\frac{1}{2}}^n := \dfrac{1}{|V|} \displaystyle\int_{V^-} (1 + \varepsilon\phi(v\sigma_{i+\frac{1}{2}})) f_{i+1}^n + \dfrac{1}{|V|} \int_{V^+} (1 + \varepsilon\phi(v\sigma_{i+\frac{1}{2}})) f_i^n, \\[3mm] \delta^L \mathcal{T}^1 f_{i+\frac{1}{2}}^n := \dfrac{\mathcal{T}^1 f_{i+\frac{1}{2}}^n - \mathcal{T}^1 f_i^n}{\Delta x/2}, \\[3mm] \delta^R \mathcal{T}^1 f_{i+\frac{1}{2}}^n := \dfrac{\mathcal{T}^1 f_{i+1}^n - \mathcal{T}^1 f_{i+\frac{1}{2}}^n}{\Delta x/2}, \end{cases}$$

with $V^+ = V \cap \mathbb{R}^+$ and $V^- = V \cap \mathbb{R}^-$. Substituting the above approximations into equation (3.21) yields an expression for $f(x_{i+\frac{1}{2}}, v, t)$ such that for $v > 0$,

$$f(x_{i+\frac{1}{2}}, v, t) = f_i^n \exp\left(-\frac{(1 + \varepsilon\phi(v\sigma_{i+\frac{1}{2}}))}{\varepsilon^2}(t - t^n)\right) + \frac{\mathcal{T}^1 f_{i+\frac{1}{2}}^n}{1 + \varepsilon\phi(v\sigma_{i+\frac{1}{2}})}$$

$$\times \left(1 - \exp\left(-\frac{(1 + \varepsilon\phi(v\sigma_{i+\frac{1}{2}}))}{\varepsilon^2}(t - t^n)\right)\right) + v\varepsilon \frac{\delta^L \mathcal{T}^1 f_{i+\frac{1}{2}}^n}{(1 + \varepsilon\phi(v\sigma_{i+\frac{1}{2}}))^2}$$

$$\times \left[\left(1 + \frac{1 + \varepsilon\phi(v\sigma_{i+\frac{1}{2}})}{\varepsilon^2}(t - t^n)\right) \exp\left(-\frac{(1 + \varepsilon\phi(v\sigma_{i+\frac{1}{2}}))}{\varepsilon^2}(t - t^n)\right) - 1\right],$$

$$\tag{3.22}$$

and for $v < 0$,

$$f(x_{i+\frac{1}{2}}, v, t) = f_{i+1}^n \exp\left(-\frac{(1 + \varepsilon\phi(v\sigma_{i+\frac{1}{2}}))}{\varepsilon^2}(t - t^n)\right) + \frac{\mathcal{T}^1 f_{i+\frac{1}{2}}^n}{1 + \varepsilon\phi(v\sigma_{i+\frac{1}{2}})}$$

$$\times \left(1 - \exp\left(-\frac{(1 + \varepsilon\phi(v\sigma_{i+\frac{1}{2}}))}{\varepsilon^2}(t - t^n)\right)\right) + v\varepsilon \frac{\delta^R \mathcal{T}^1 f_{i+\frac{1}{2}}^n}{(1 + \varepsilon\phi(v\sigma_{i+\frac{1}{2}}))^2}$$

$$\times \left[\left(1 + \frac{1 + \varepsilon\phi(v\sigma_{i+\frac{1}{2}})}{\varepsilon^2}(t - t^n)\right) \exp\left(-\frac{(1 + \varepsilon\phi(v\sigma_{i+\frac{1}{2}}))}{\varepsilon^2}(t - t^n)\right) - 1\right].$$

$$\tag{3.23}$$

- **Determine** $\Phi^n_{i+\frac{1}{2}}, F^n_{i+\frac{1}{2}}$. The flux $\Phi^n_{i+\frac{1}{2}}(v)$ in (3.19) can be approximated by

$$
\begin{aligned}
\Phi_{i+\frac{1}{2}}(v) &= Avf^n_{i+1} + BvT^1 f^n_{i+\frac{1}{2}} + Cv^2\delta^R T^1 f^n_{i+\frac{1}{2}}, \quad \text{for } v < 0, \\
\Phi_{i+\frac{1}{2}}(v) &= Avf^n_i + BvT^1 f^n_{i+\frac{1}{2}} + Cv^2\delta^L T^1 f^n_{i+\frac{1}{2}}, \quad \text{for } v > 0,
\end{aligned}
\tag{3.24}
$$

where the coefficients $A(v, \varepsilon, \Delta t), B(v, \varepsilon, \Delta t)$, and $C(v, \varepsilon, \Delta t)$ can be determined explicitly, such that

$$
A(v, \varepsilon, \Delta t) := \frac{\varepsilon}{\Delta t(1 + \varepsilon\phi(v\sigma_{i+\frac{1}{2}}))} \left(1 - \exp\left(-\frac{1 + \varepsilon\phi(v\sigma_{i+\frac{1}{2}})}{\varepsilon^2}\Delta t \right) \right),
$$

$$
B(v, \varepsilon, \Delta t) := \frac{1}{\varepsilon(1 + \varepsilon\phi(v\sigma_{i+\frac{1}{2}}))}
$$
$$
- \frac{\varepsilon}{\Delta t(1 + \varepsilon\phi(v\sigma_{i+\frac{1}{2}}))^2} \left(1 - \exp\left(-\frac{1 + \varepsilon\phi(v\sigma_{i+\frac{1}{2}})}{\varepsilon^2}\Delta t \right) \right),
$$

$$
C(v, \varepsilon, \Delta t) := \frac{2\varepsilon^2}{\Delta t(1 + \varepsilon\phi(v\sigma_{i+\frac{1}{2}}))^3} \left(1 - \exp\left(-\frac{1 + \varepsilon\phi(v\sigma_{i+\frac{1}{2}})}{\varepsilon^2}\Delta t \right) \right)
$$
$$
- \frac{1}{(1 + \varepsilon\phi(v\sigma_{i+\frac{1}{2}}))^2} \left(1 + \exp\left(-\frac{1 + \varepsilon\phi(v\sigma_{i+\frac{1}{2}})}{\varepsilon^2}\Delta t \right) \right).
\tag{3.25}
$$

Furthermore, $F^n_{i+\frac{1}{2}}$ in (3.19) is given by

$$
\begin{aligned}
F^n_{i+\frac{1}{2}} &= \frac{1}{|V|}\int_{V-} Avf^n_{i+1}dv + \frac{1}{|V|}\int_{V+} Avf^n_i dv + \frac{1}{|V|}T^1 f^n_{i+\frac{1}{2}}\int_V vBdv \\
&+ \frac{1}{|V|}\delta^R T^1 f^n_{i+\frac{1}{2}}\int_{V-} Cv^2 dv + \frac{1}{|V|}\delta^L T^1 f^n_{i+\frac{1}{2}}\int_{V+} Cv^2 dv.
\end{aligned}
\tag{3.26}
$$

This concludes the construction of the scheme.

3.3.3 SP property

The UGKS scheme has to meet the two requirements of the SP property. The AP discretization of the steady state equation is given in Appendix C. For the second requirement, we assume that we start from a steady state solution, that at the discrete level satisfies,

$$
\frac{\Phi^n_{i+\frac{1}{2}} - \Phi^n_{i-\frac{1}{2}}}{\Delta x} = \frac{1}{\varepsilon^2}(\rho^n_i - f^n_i) + \frac{1}{\varepsilon}\left(\frac{1}{|V|}\int_V \phi(v'\sigma_{i+\frac{1}{2}})f^n_i(v')\,dv' - \phi(v\sigma_{i+\frac{1}{2}})f^n_i \right).
\tag{3.27}
$$

Integrating equation (3.27) over v yields

$$
\frac{F^n_{i+\frac{1}{2}} - F^n_{i-\frac{1}{2}}}{\Delta x} = 0.
$$

From (3.17) one can deduce that,

$$\rho_i^{n+1} = \rho_i^n, \tag{3.28}$$

which indicates that the macroscopic density is preserved. Using (3.27), the equation of updating f^{n+1} in (3.18) can be written as,

$$\frac{f_i^{n+1} - f_i^n}{\Delta t} = \frac{1}{\varepsilon^2}\left((\rho_i^{n+1} - \rho_i^n) - (f_i^{n+1} - f_i^n)\right).$$

Then from (3.28),

$$\left(1 + \frac{\Delta t}{\epsilon^2}\right)(f_i^{n+1} - f_i^n) = 0,$$

which gives $f_i^{n+1} = f_i^n$. This concludes the SP property of the UGKS.

3.3.4 Numerical results

Parameters in equation (3.13) are chosen as in Gosse [33] such that,

$$\chi_S = 1, D = 15, \beta = 60, \alpha = 3.$$

and ϕ is of the form

$$\phi(u) = -\chi_S \tanh u.$$

The computational domain is set to be $x \in [-1, 1]$. We impose specular boundary conditions for f and Dirichlet conditions for S. The initial density distribution is composed of two bumps located at $x = \pm 0.65$ given by:

$$f(x, v, 0) = 5(\exp(-10(x-0.65)^2 - 20(v+0.45)^2) + \exp(-10(x+0.65)^2 - 20(v-0.45)^2)).$$

We use $\Delta x = 2/500$ for the space discretization and $v \in [-1, 1]$ with the S_{32} Gaussian quadrature points for the velocity. The limiting scheme of the UGKS is an explicit solver for the diffusion equation. Therefore, to ensure the stability of the numerical scheme, the time step Δt is chosen as below

$$\Delta t = \begin{cases} 0.5\Delta x^2, & \text{for } \varepsilon < \Delta x, \\ 0.5\varepsilon\Delta x, & \text{else.} \end{cases}$$

In order to verify the AP property of our scheme, the total densities ρ at time $t = 1$ are displayed in figure 3.3 for different values of ε ranging from 10^{-2} to 10^{-6}. In order to check the SP property, we give the time evolution of the L^∞-norm of the difference between the solution and the stationary solution in the time interval $[0,100]$ in Table 3.3 for $\varepsilon = 1$ and $\varepsilon = 10^{-3}$. These results ensure that the SP property is independent of ε.

T	0	30	60	65	100
L^∞	0.9064	8.260×10^{-7}	3.767×10^{-11}	7.474×10^{-12}	1.662×10^{-12}
T	0	5	10	50	100
L^∞	0.6493	3.024×10^{-7}	2.064×10^{-9}	2.199×10^{-10}	1.476×10^{-10}

Table 3.3: Chemotaxis: L^∞-norm of the difference between the solution and the stationary solution in the time interval [0,100] for $\varepsilon = 1$ (top) and $\varepsilon = 10^{-3}$ (bottom).

3.4 IMEX scheme with the Penalization method for the Boltzmann equation

In this section, we consider the penalization method developed in [29] for the Boltzmann equation. This method together with an IMEX discretization of the equation give an AP scheme for the Boltzmann equation. One can find the AP proof in [29]. Here we show that the penalization method is not only AP but also SP. In [25], the authors propose a multistep high order IMEX AP scheme for the BGK model and the Boltzmann equation. The scheme is originally developed for the BGK model and then extended by the penalization method to the Boltzmann equation. One can think of the scheme proposed in [25] as the high order version of the scheme in [29]. The authors prove that the IMEX AP scheme, without penalization, is SP uniformly in ε. Our criterion can be applied successfully to the high order IMEX AP scheme in [25] after penalization. Our proof, in contrast to theirs, requires the linear dependence of the Maxwellian of the collision operator.

3.4.1 The Boltzmann equation

The Boltzmann equation describes the time evolution of the density distribution of gas particles. It is given by

$$\partial_t f + v \cdot \nabla_x f = \frac{\mathcal{Q}(f)}{\varepsilon}.$$

Here $f(x, v, t)$ is the probability density distribution of particles at time t, position x and with velocity v. $\mathcal{Q}$ is the Boltzmann collision operator where only binary interactions are considered. Let (v, v_*) and (v', v'_*) be respectively the velocities of the two colliding particles before and after the collision related by

$$\begin{cases} v' = \frac{1}{2}((v - v_*) - |v - v_*|\sigma), \\ v'_* = \frac{1}{2}((v - v_*) + |v - v_*|\sigma). \end{cases}$$

With $\sigma \in \mathbb{S}^{d_v-1}$. $\mathcal{Q}$ is given by

$$\mathcal{Q}(f)(v) = \int_{\mathbb{R}^{d_v}} \int_{\mathbb{S}^{d_v-1}} B(|v - v_*|, \cos\theta)(f(v'_*)f(v') - f(v_*)f(v))d\sigma dv_*. \quad (3.29)$$

The collision kernel B is a non-negative function given by $B(|u|, \cos\theta) = C_\lambda |u|^\lambda$, where $u = \frac{(v - v_*)}{|v - v_*|}$ and $\cos\theta = u \cdot \sigma$, for some $\lambda \in [0, 1]$ and a constant $C_\lambda > 0$. For more details, one can look at the Boltzmann equation description in [29]. ε is the dimensionless Knudsen number and $\int_v \omega(v)Q(f)dv = 0$ for $\omega(v) = (1, v, |v|^2)$. The equilibrium distribution of Q is the Maxwellian distribution $\mathcal{M}_{\rho,u,T}$, i.e. $Q(\mathcal{M}_{\rho,u,T}) = 0$ and it is given by,

$$\mathcal{M}_{\rho,u,T}(v) = \frac{\rho}{(2\pi T)^{\frac{d_v}{2}}} \frac{1}{\exp \frac{|v-u|^2}{2T}},$$

where ρ, u, and T are the density, velocity and temperature of the gas, and d_v is the dimension of the velocity space. As $\varepsilon \to 0$, the zeroth, first and second moments of the distribution function solve the Euler equations.

3.4.2 IMEX scheme with the penalization method

The penalization method was originally developed in [29, 47]. The purpose is to split the collision term of the Boltzmann equation into a stiff part and less stiff part. More precisely, the Boltzmann equation is written in the following form:

$$\partial_t f + v \cdot \nabla_x f = \frac{Q(f) - P(f)}{\varepsilon} + \frac{P(f)}{\varepsilon},$$

where $Q(f)$ is the Boltzmann collision operator and $P(f)$ is a relaxation operator, namely $P(f) = \beta[\mathcal{M}_{\rho,u,T}(v) - f(v)]$, where β is a strictly positive parameter. $P(f)$ has the same equilibrium as $Q(f)$. It satisfies $\int_v P(f)\omega(v)dv = 0$ for $\omega(v) = (1, v, |v|^2)$ and $P(\mathcal{M}_{\rho,u,T}) = 0$. As in [29], β^n is chosen to be $2\pi\rho^n$ such that both operators $P(f)$ and the full Boltzmann operator $Q(f)$ have the same loss term corresponding to the dissipative part.

The following IMEX discretization of the Boltzmann equation is proposed in [29]:

$$\frac{f^{n+1} - f^n}{\Delta t} + v \cdot \nabla_x f^n = \frac{Q(f^n) - P(f^n)}{\varepsilon} + \frac{P(f^{n+1})}{\varepsilon}. \tag{3.30}$$

For the discretization of the Boltzmann operator, one can use a fast spectral Fourier-Galerkin method [30], and for the transport part, a first or second order finite volume scheme can be employed. This gives an AP discretization for the Boltzmann equation as proven in [29].

3.4.3 SP property

Because we computed our numerical results in a space homogeneous set up, proving that the discretization of the steady state equation is AP, is unnecessary, knowing that the full scheme is AP [29]. We only need to prove that starting from a discrete stationary solution, the solution of the time evolutionary problem does not change. Suppose that the solution satisfies the stationary equation at time t^n, i.e.

$$v \cdot \nabla_x f^n = \frac{Q(f^n) - P(f^n)}{\varepsilon} + \frac{P(f^n)}{\varepsilon}. \tag{3.31}$$

It follows from the properties of the collision operator $\mathcal{Q}$ and the relaxation operator P that:

$$\int_v \omega(v)v \cdot \nabla_x f^n = 0, \tag{3.32}$$

with $\omega(v) = (1, v, |v|^2)$.

Now multiply (3.30) by $\omega(v)$ and integrate over the velocity space. Using the conservation properties of $\mathcal{Q}$, P and (3.32), one observes that the Maxwellian of the distribution function is preserved. Substituting (3.31) in (3.30) gives,

$$\frac{f^{n+1} - f^n}{\Delta t} = \frac{-P(f^n)}{\varepsilon} + \frac{P(f^{n+1})}{\varepsilon}.$$

Now, we plug in P by its defnition $P(f) = \beta[\mathcal{M}_{\rho,u,T}(v) - f(v)]$,

$$\frac{f^{n+1} - f^n}{\Delta t} = -\frac{\beta^n[\mathcal{M}^n - f^n]}{\varepsilon} + \frac{\beta^{n+1}[\mathcal{M}^{n+1} - f^{n+1}]}{\varepsilon}.$$

Since $\mathcal{M}^{n+1} = \mathcal{M}^n$ and $\beta^{n+1} = \beta^n$, $f^{n+1} = f^n$ and the steady state is preserved.

3.4.4 Numerical results

In this section, we consider the 2D Bose gas experiment 3.3 in [41] to test the AP and the SP property of the penalization method presented in [29]. We solve the space homogeneous quantum Boltzmann equation in 2D velocity space which is a special case of the classical Boltzmann equation for a particular collision operator $\mathcal{Q}_q$.

$$\partial_t f = \frac{\mathcal{Q}_q(f) - P(f)}{\varepsilon} + \frac{P(f)}{\varepsilon}.$$

As defined in [41], the quantum collision operator is another version of the collision operator (3.29) and given by

$$\mathcal{Q}_q(f)(v) = \int_{\mathbb{R}^{d_v}} \int_{\mathbb{S}^{d_v-1}} B(|v - v_*|, \cos\theta) \Big(f'_* f'(1 \pm \theta_0 f)(1 \pm \theta_0 f_*)$$
$$- f_* f(1 \pm \theta_0 f')(1 \pm \theta_0 f'_*) \Big) d\sigma dv_*$$

where $\theta_0 = h^{d_v}$ and h is the rescaled Planck constant. The upper sign corresponds to the Bose gas, while the lower sign to the Fermi gas. In this experiment we consider the Bose gas case. The idea can be extended to more general collision operators. Hence, scheme (3.30) is simplified to

$$f^{n+1} = \frac{\varepsilon}{\varepsilon + \beta^{n+1}\Delta t} f^n + \Delta t \frac{\mathcal{Q}_q(f^n) - P(f^n)}{\varepsilon + \beta^{n+1}\Delta t} + \frac{\beta^{n+1}\Delta t}{\varepsilon + \beta^{n+1}\Delta t} \mathcal{M}^{n+1}.$$

The initial distribution function is given as in [11],

$$f_0(v) = \frac{\rho_0}{4\pi T_0} \left(\exp\left(\frac{-|v - u_0|^2}{2T_0} \right) + \exp\left(\frac{-|v + u_0|^2}{2T_0} \right) \right),$$

where $\rho_0 = 1$, $T_0 = 3/8$, and $u_0 = (1, 1/2)$. The computational domain is $[-8, 8]^2$ with 64 grid points. The quantum Maxwellian [41] is given as,

$$\mathcal{M}_q(v) = \frac{1}{\theta_0} \frac{1}{z^{-1} \exp \frac{|v-u|^2}{2T} - 1},$$

where $\theta_0 = 0.1^2$, $z = 0.001590$, $T = 1$ is the temperature, and $u = 0$ is the macroscopic velocity. In figure 3.4, we test the AP property of the penalization method. A cross section of the distribution function for different values of ε is plotted on the left and a zoomed part of the plot on the right. The curves get closer to each other as ε converges to 0 which implies the AP property. Next, we investigate the SP property.

Figure 3.5 shows contours of the 2D distribution function and the contour lines of the difference between the distribution function f and its equilibrium at $t = 200$. We computed the L^∞-norm of the difference between f and its equilibrium in the time interval $[0, 200]$ in figure 3.6 as evidence that f converges exponentially to the equilibrium. Table 3.4 presents the L^∞ norm of the distances between the time evolutionary simulation and the equilibrium at some discrete times, where one can find exactly when the initial distribution function reaches its equilibrium.

T	0	20	50	100	150	200
L^∞	0.545	1.2×10^{-3}	6.58×10^{-7}	3.49×10^{-12}	7.61×10^{-13}	5.62×10^{-13}

Table 3.4: Boltzmann: L^∞-norm of the difference between f and its equilibrium starting from t=0 until the final time t=200 for $\varepsilon = 1$.

3.5 Conclusion

Proving the SP property for the three AP schemes, leads neadly to a criterion, emphasising that AP schemes with a discretization that linearly depends on the Maxwellian are also SP [27]. We realized that the linear dependency on the Maxwellian in the source term is the key to proving that the moments are being updated explicitly not implicitly. This in turn is the key to proving ultimately that the updated solution at the next time t^{n+1} does not change in the case of steady state solutions.

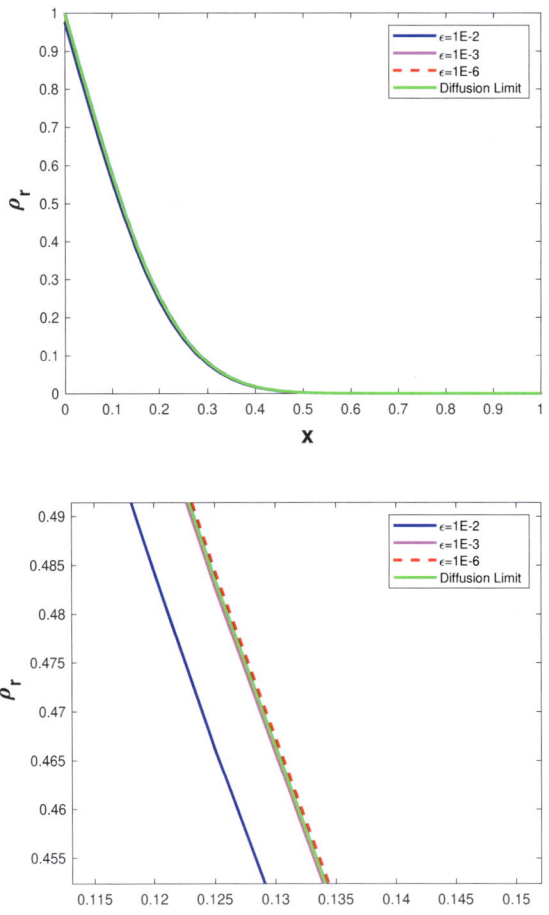

Figure 3.1: Neutron Transport: Left: the density ρ_r at time $t = 0.05$ for $\varepsilon = 10^{-2}, \varepsilon = 10^{-3}, \varepsilon = 10^{-6}$ and the solution of the diffusion limit equation; right: a zoomed part of the left plot.

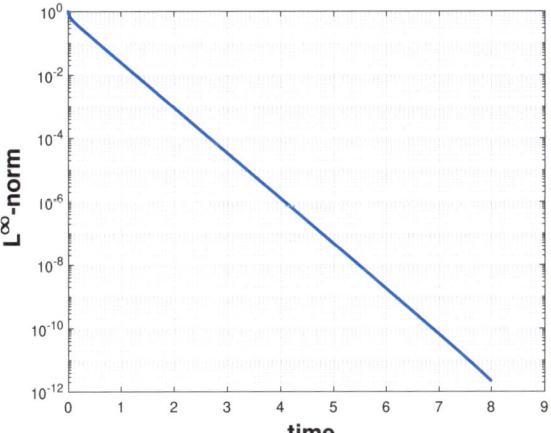

Figure 3.2: Neutron Transport: time evolution of the L^∞-norm of the difference between the solution and the stationary solution in the time interval [0,8] for $\varepsilon = 10^{-8}$.

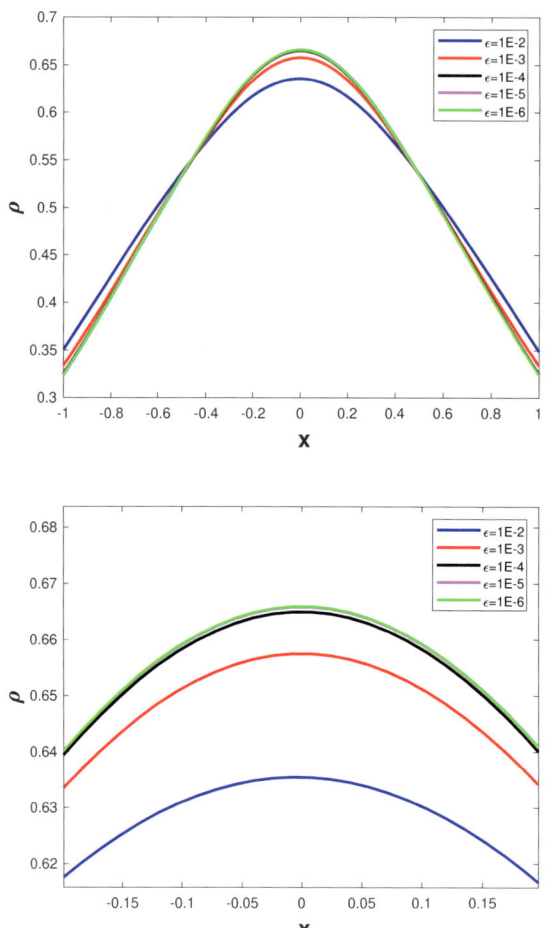

Figure 3.3: Chemotaxis: left, the density ρ at time $t = 1$ for $\varepsilon = 10^{-2}, 10^{-3}, 10^{-4}, 10^{-5}, 10^{-6}$; right, a zoomed part of the left plot.

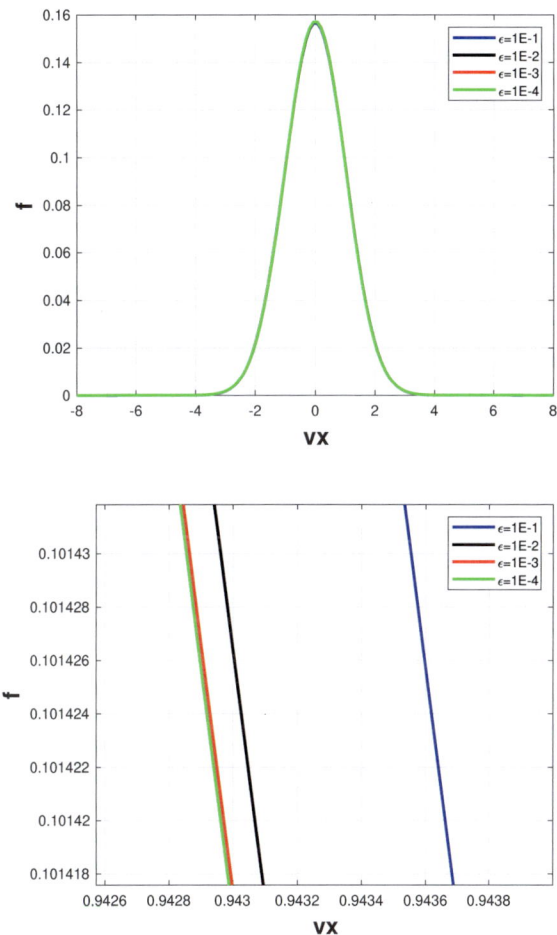

Figure 3.4: Boltzmann: cross section of the distribution function for different values of ε(left) and a zoomed part of the plot(right).

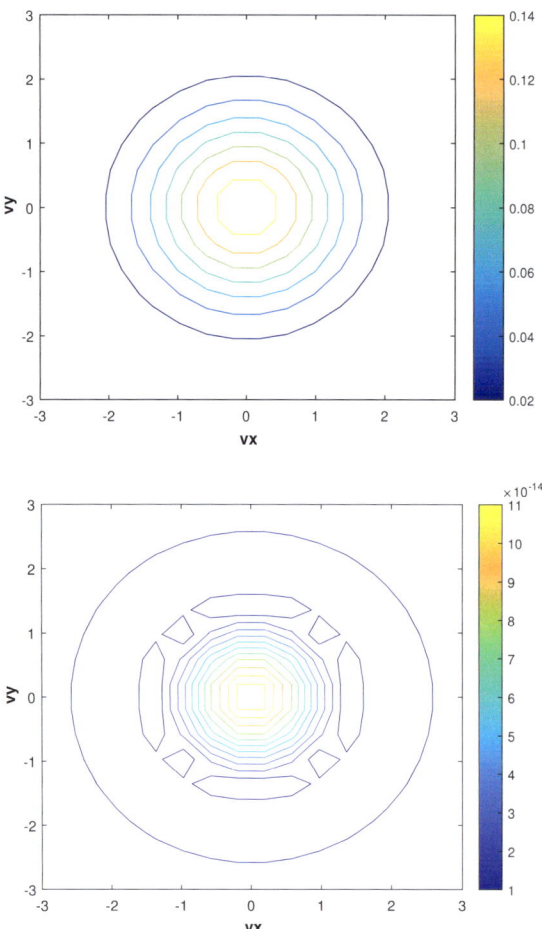

Figure 3.5: Boltzmann: contours of the 2D distribution function (top) and the contour lines of the difference between the distribution function and its equilibrium (bottom) at the final time $t = 200$.

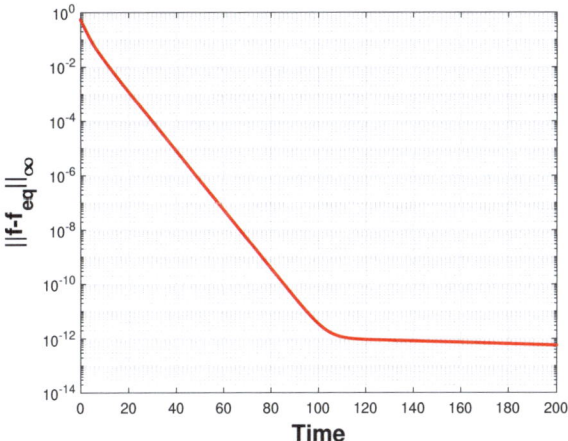

Figure 3.6: Boltzmann: time evolution of the L^{∞}-norm of the difference between the distribution function f and its equilibrium in the time interval $[0,200]$.

Chapter 4

AP and SP Schemes for the Isentropic Euler Equations with Gravity

4.1 Introduction

The resulted criterion at the kinetic level causes us to consider how to translate this to the fluid level. For this reason, we consider again a fluid model in this section and we try to develop an AP scheme and investigate the relationship between AP and SP properties for fluid models [50]. We start with a special case of the Euler system, namely the isentropic Euler system. We extend the AP scheme developed by Goudon et al. for the isentropic Euler equations to the case with gravitational source term. In section 4.2, we introduce the model. We present the AP and SP semi-discrete numerical scheme in section 4.3 and the fully-dicrete scheme in the 1D and 2D framework in section 4.4. Some numerical test cases from the litrature to validate the properties of the scheme are considered in section 4.5.

4.2 The Isentropic Euler Equations with Gravity

4.2.1 The model

The Isentropic Euler equations with gravitational source term is a special case of the Euler equations (2.49) and is given by,

$$\partial_t \rho + \nabla.(\rho \mathbf{u}) = 0,$$
$$\partial_t(\rho \mathbf{u}) + \nabla.(\rho \mathbf{u} \otimes \mathbf{u}) + \nabla p(\rho) = -\rho \nabla \phi. \tag{4.1}$$

Where ρ is the density, $\mathbf{u}$ is the velocity field, p is the pressure, and $\rho \mathbf{u}$ is the momentum. The pressure law is given by $p(\rho) = A\rho^\gamma$, where A and γ are positive constants. ϕ is the gravitational potential, a given function of space.

4.2.2 Scaling

One scales the equations (4.1) to describe the low Mach number (incompressible) limit. Let $x_0, t_0, \rho_0, p_0, u_0$ be a set of characteristic scales for the variables. The dimensionless variables are then given by, $\hat{x} = \frac{x}{x_0}, \hat{t} = \frac{t}{t_0}, \hat{\phi} = \frac{\phi}{\phi_0}, \dots$ with $\phi_0 = \frac{p_0}{\rho_0}$.

Substitute the variables in the equations,

$$\frac{1}{t_0}\partial_{\hat{t}}(\hat{\rho}\rho_0) + \frac{1}{x_0}\nabla_{\cdot\hat{x}}(\hat{\rho}\rho_0\hat{\mathbf{u}}\mathbf{u}_0) = 0,$$

$$\frac{1}{t_0}\partial_{\hat{t}}(\hat{\rho}\rho_0\hat{\mathbf{u}}\mathbf{u}_0) + \frac{1}{x_0}\nabla_{\cdot\hat{x}}(\hat{\rho}\rho_0\hat{\mathbf{u}}\mathbf{u}_0 \otimes \hat{\mathbf{u}}\mathbf{u}_0) + \frac{1}{x_0}\nabla_{\hat{x}}(\hat{p}p_0) = -(\hat{\rho}\rho_0)\frac{1}{x_0}\nabla_{\cdot\hat{x}}(\hat{\phi}\phi_0). \tag{4.2}$$

then,

$$\frac{\rho_0}{t_0}\partial_{\hat{t}}\hat{\rho} + \frac{\rho_0\mathbf{u}_0}{x_0}\nabla_{\cdot\hat{x}}(\hat{\rho}\hat{\mathbf{u}}) = 0,$$

$$\frac{\rho_0\mathbf{u}_0}{t_0}\partial_{\hat{t}}(\hat{\rho}\hat{\mathbf{u}}) + \frac{\rho_0\mathbf{u}_0^2}{x_0}\nabla_{\cdot\hat{x}}(\hat{\rho}\hat{\mathbf{u}} \otimes \hat{\mathbf{u}}) + \frac{p_0}{x_0}\nabla_{\hat{x}}(\hat{p}) = -\frac{\rho_0\phi_0}{x_0}\hat{\rho}\nabla_{\cdot\hat{x}}\hat{\phi}. \tag{4.3}$$

Drop the hat,

$$\frac{\rho_0}{t_0}\partial_t\rho + \frac{\rho_0\mathbf{u}_0}{x_0}\nabla_{\cdot}(\rho\mathbf{u}) = 0,$$

$$\frac{\rho_0\mathbf{u}_0}{t_0}\partial_t(\rho\mathbf{u}) + \frac{\rho_0\mathbf{u}_0^2}{x_0}\nabla_{\cdot}(\rho\mathbf{u} \otimes \mathbf{u}) + \frac{p_0}{x_0}\nabla p = -\frac{\rho_0\phi_0}{x_0}\rho\nabla\phi. \tag{4.4}$$

Multiply by $\frac{t_0}{\rho_0}$ and use the fact that $\mathbf{u}_0 = \frac{x_0}{t_0}$,

$$\partial_t\rho + \nabla_{\cdot}(\rho\mathbf{u}) = 0,$$

$$\mathbf{u}_0\partial_t(\rho\mathbf{u}) + \mathbf{u}_0\nabla_{\cdot}(\rho\mathbf{u} \otimes \mathbf{u}) + \frac{p_0}{\rho_0\mathbf{u}_0}\nabla p = -\frac{\phi_0}{\mathbf{u}_0}\rho\nabla\phi. \tag{4.5}$$

So,

$$\partial_t\rho + \nabla_{\cdot}(\rho\mathbf{u}) = 0,$$

$$\partial_t(\rho\mathbf{u}) + \nabla_{\cdot}(\rho\mathbf{u} \otimes \mathbf{u}) + \frac{p_0}{\rho_0\mathbf{u}_0^2}\nabla p = -\frac{\phi_0}{\mathbf{u}_0^2}\rho\nabla\phi. \tag{4.6}$$

Then,

$$\partial_t\rho + \nabla_{\cdot}(\rho\mathbf{u}) = 0,$$

$$\partial_t(\rho\mathbf{u}) + \nabla_{\cdot}(\rho\mathbf{u} \otimes \mathbf{u}) + \frac{1}{\varepsilon^2}\nabla p(\rho) = -\frac{1}{\varepsilon^2}\rho\nabla\phi, \tag{4.7}$$

are the non-dimensionalized equations with $\varepsilon^2 = \frac{\rho_0\mathbf{u}_0^2}{P_0}$. Taking the gravitational source term to the left-hand side and using the pressure law, system (4.7) is then,

$$\partial_t\rho + \nabla_{\cdot}(\rho\mathbf{u}) = 0,$$

$$\partial_t(\rho\mathbf{u}) + \nabla_{\cdot}(\rho\mathbf{u} \otimes \mathbf{u}) + \frac{1}{\varepsilon^2}\rho\nabla W = 0, \tag{4.8}$$

with

$$W = \frac{A\gamma}{\gamma - 1}\rho^{\gamma-1} + \phi. \tag{4.9}$$

4.2.3 The incompressible limit equations

Assume that the Chapman-Enskog asymptotic expansions of the variables are,

$$\rho = \rho^{(0)} + \varepsilon\rho^{(1)} + \varepsilon^2\rho^{(2)} + \cdots$$

$$\mathbf{u} = \mathbf{u}^{(0)} + \varepsilon\mathbf{u}^{(1)} + \varepsilon^2\mathbf{u}^{(2)} + \cdots$$

$$W = W^{(0)} + \varepsilon W^{(1)} + \varepsilon^2 W^{(2)} + \cdots \quad .$$

The expansion of $\rho\nabla W$ can be seen as,

$$\rho\nabla W = (\rho^{(0)} + \varepsilon\rho^{(1)} + \varepsilon^2\rho^{(2)} + \cdots)\nabla(W^{(0)} + \varepsilon W^{(1)} + \varepsilon^2 W^{(2)} + \cdots),$$

$$= \rho^{(0)}\nabla W^{(0)} + \varepsilon\Big(\rho^{(0)}\nabla W^{(1)} + \rho^{(1)}\nabla W^{(0)}\Big)$$

$$+ \varepsilon^2\Big(\rho^{(0)}\nabla W^{(2)} + \rho^{(1)}\nabla W^{(1)} + \rho^{(2)}\nabla W^{(0)}\Big) + \cdots. \quad (4.10)$$

Comparing the $\mathcal{O}(\frac{1}{\varepsilon^2})$ terms in system (4.8) and using $\rho^{(0)} \neq 0$, one deduces that $\nabla W^{(0)} = 0$. Then looking at the $\mathcal{O}(\frac{1}{\varepsilon})$ terms yields $\nabla W^{(1)} = 0$. From the definition of W in (4.9), $\nabla W^{(0)} = 0$ leads to,

$$\frac{A\gamma}{\gamma-1}\big(\rho^{(0)}\big)^{\gamma-1} + \phi(x) = c$$

with c being a constant independent of x.

$$\big(\rho^{(0)}\big)^{\gamma-1} + \frac{\gamma-1}{A\gamma}\phi(x) = c'$$

with c' being a constant independent of x. c' is proved to be 1 by the boundary conditions of ρ.

$$\rho^{(0)} = \left(1 - \frac{\gamma-1}{\gamma A}\phi(x)\right)^{\frac{1}{\gamma-1}}, \quad (4.11)$$

which indicates that when $\varepsilon \ll 1$, $\rho^{(0)}$ becomes stationary. To find the equation that $\mathbf{u}^{(0)}$ satisfies, we consider the $\mathcal{O}(1)$ terms in system (4.8) such that

$$\partial_t\rho^{(0)} + \nabla \cdot (\rho^{(0)}\mathbf{u}^{(0)}) = 0,$$

$$\partial_t(\rho^{(0)}\mathbf{u}^{(0)}) + \nabla \cdot (\rho^{(0)}\mathbf{u}^{(0)} \otimes \mathbf{u}^{(0)}) + \rho^{(0)}\nabla W^{(2)} + \rho^{(1)}\nabla W^{(1)} + \rho^{(2)}\nabla W^{(0)} = 0. \quad (4.12)$$

Using $\nabla W^{(0)} = \nabla W^{(1)} = 0$, system (4.12) can be written as

$$\nabla \cdot (\rho^{(0)}\mathbf{u}^{(0)}) = 0,$$
$$\partial_t\mathbf{u}^{(0)} + \mathbf{u}^{(0)}\nabla \cdot \mathbf{u}^{(0)} + \nabla W^{(2)} = 0. \quad (4.13)$$

(4.13) is the incompressible isentropic Euler equations.

4.3 Semi-discrete Numerical Scheme

4.3.1 The scheme

Following the splitting technique introduced in [37] and used in [34], we split the divergence in the density equation as well as the pressure and the gravitational source term in the momentum equation. Let

$$\rho_0 = \left(1 - \frac{\gamma - 1}{\gamma A}\phi(x)\right)^{\frac{1}{\gamma - 1}}.$$

System (4.7) can be written as:

$$\partial_t \rho + \alpha \nabla \cdot (\rho \mathbf{u}) + (1 - \alpha)\nabla \cdot (\rho \mathbf{u}) = 0,$$

$$\partial_t (\rho \mathbf{u}) + \nabla \cdot (\rho \mathbf{u} \otimes \mathbf{u}) + \frac{1}{\varepsilon^2}\nabla \left[p(\rho) - a(t)\rho\right] + \frac{1}{\varepsilon^2}a(t)\nabla \rho \qquad (4.14)$$

$$= -\frac{1}{\varepsilon^2}\rho\nabla\phi - \frac{a(t)}{\varepsilon^2}\rho\nabla \ln \rho_0 + \frac{a(t)}{\varepsilon^2}\rho\nabla \ln \rho_0.$$

We will see the necessity of this modification in the AP proof later on. The system is splitted into the following two subsystems:

$$\partial_t \rho + \alpha \nabla \cdot (\rho \mathbf{u}) = 0,$$

$$\partial_t (\rho \mathbf{u}) + \nabla \cdot (\rho \mathbf{u} \otimes \mathbf{u}) + \frac{1}{\varepsilon^2}\nabla \left[p(\rho) - a(t)\rho\right] = -\frac{1}{\varepsilon^2}\rho\nabla\phi - \frac{a(t)}{\varepsilon^2}\rho\nabla \ln \rho_0, \qquad (4.15)$$

and

$$\partial_t \rho + (1 - \alpha)\nabla \cdot (\rho \mathbf{u}) = 0,$$

$$\partial_t (\rho \mathbf{u}) + \frac{1}{\varepsilon^2}a(t)\nabla \rho = \frac{a(t)}{\varepsilon^2}\rho\nabla \ln \rho_0, \qquad (4.16)$$

where $0 \le \alpha < 1$ is a constant and the time dependent function $a(t) > 0$ depends on the hyperbolicity of the system (4.15). The first system (4.15) takes into account the slow speed, in the conservative form, it can be written as

$$U_t + F(U)_x + G(U)_y = S(U), \qquad (4.17)$$

with

$$U = \begin{pmatrix} \rho \\ \rho u \\ \rho v \end{pmatrix}, \qquad F(U) = \begin{pmatrix} \alpha \rho u \\ \rho u^2 + \frac{p(\rho) - a(t)\rho}{\varepsilon^2} \\ \rho uv \end{pmatrix},$$

$$G(U) = \begin{pmatrix} \alpha \rho v \\ \rho uv \\ \rho v^2 + \frac{p(\rho) - a(t)\rho}{\varepsilon^2} \end{pmatrix}, \qquad S(U) = \begin{pmatrix} 0 \\ -\frac{1}{\varepsilon^2}\rho\phi_x - \frac{a(t)}{\varepsilon^2}\rho(\ln \rho_0)_x \\ -\frac{1}{\varepsilon^2}\rho\phi_y - \frac{a(t)}{\varepsilon^2}\rho(\ln \rho_0)_y \end{pmatrix},$$

where $\mathbf{u} = (u, v)^T$. The eigenvalues of the jacobian matrix of $F(U)$ are

$$\lambda_1 = u, \quad \lambda_2 = u + c, \quad \lambda_3 = u - c, \quad \text{with} \quad c(\rho, u) = \sqrt{(1 - \alpha)u^2 + \alpha \frac{p'(\rho) - a(t)}{\varepsilon^2}}.$$

Similar calculations hold for $G(U)$, the flux along y. The choice of $a(t)$ is to guarantee the hyperbolicity of the system (4.15). We choose $a(t)$ such that the eigenvalues of the jacobian matrices of $F(U)$, $G(U)$ are real and positive i.e. $p'(\rho) \geq a(t)$.

Hence, $a(t)$ is chosen as the following, $a(t) = \min_x \{p'(\rho)\}$. But, with this choice, spurious oscillations are observed in some test cases for large values of ε. They appear in regions where the density is nearly uniform and the material velocity vanishes. Indeed, in these regions, the corresponding sound speed vanishes and the spurious oscillations are probably due to a lack of numerical diffusion in the Slow Dynamic part of the splitting. Now, we add the function $l(t)$ that eliminates the spurious oscillations that might appear for large ε. Thus, $a(t) = \min_x \{p'(\rho)\} - l(t)\varepsilon^2$, $l(t)$ is a constant such that $a(t) \geq l(t)(1 - \varepsilon^2) > 0$ for $\varepsilon < 1$. In the numerical tests we choose $l(t) = 0$ or $l(t) = 1$, for more details, see [34]. Let Δt be the time step, $t^0 = 0$, and for a positive integer n, we set $t^{n+1} = t^n + \Delta t$. The two subsystems (4.15), (4.16) can now be discretized as,

$$\begin{cases} \frac{\rho^* - \rho^n}{\Delta t} + \alpha \nabla \cdot (\rho \mathbf{u})^n = 0, \\ \frac{(\rho \mathbf{u})^* - (\rho \mathbf{u})^n}{\Delta t} + \nabla \cdot (\rho \mathbf{u} \otimes \mathbf{u})^n + \frac{1}{\varepsilon^2} \nabla [p(\rho) - a(t)\rho]^n = -\frac{1}{\varepsilon^2} \rho^n \nabla \phi - \frac{a^n}{\varepsilon^2} \rho^n \nabla \ln \rho_0, \end{cases}$$
$$(4.18)$$

and

$$\begin{cases} \frac{\rho^{n+1} - \rho^*}{\Delta t} + (1 - \alpha) \nabla \cdot (\rho \mathbf{u})^{n+1} = 0, \\ \frac{(\rho \mathbf{u})^{n+1} - (\rho \mathbf{u})^*}{\Delta t} + \frac{1}{\varepsilon^2} a(t^n) \nabla \rho^{n+1} = \frac{a^n}{\varepsilon^2} \rho^{n+1} \nabla \ln \rho_0. \end{cases} \quad (4.19)$$

4.3.2 The AP property

To illustrate the idea, we start from the AP proof of the semi-discretized scheme. By substituting ρ^*, $\mathbf{u}^*$ from the first system (4.32) into the second one (4.33), the two semi-discrete systems (4.32) and (4.33) can be viewed as,

$$\frac{\rho^{n+1} - \rho^n}{\Delta t} + \alpha \nabla \cdot (\rho \mathbf{u})^n + (1 - \alpha) \nabla \cdot (\rho \mathbf{u})^{n+1} = 0,$$

$$\frac{(\rho \mathbf{u})^{n+1} - (\rho \mathbf{u})^n}{\Delta t} + \nabla \cdot (\rho \mathbf{u} \otimes \mathbf{u})^n + \frac{1}{\varepsilon^2} \nabla [p(\rho) - a\rho]^n + \frac{1}{\varepsilon^2} a^n \nabla \rho^{n+1}$$

$$= -\frac{1}{\varepsilon^2} \rho^n \nabla \phi - \frac{a^n}{\varepsilon^2} \rho^n \nabla \ln \rho_0 + \frac{a^n}{\varepsilon^2} \rho^{n+1} \nabla \ln \rho_0.$$
$$(4.20)$$

In order to prove the AP property of the semi-discrete scheme we need to prove that as ε goes to zero, (4.20) is a good discretization of the incompressible limit equation (4.13).

We reformulate the momentum equation in (4.20) before the expansion, as at the PDE level. Let

$$M(\rho) = \int_q^\rho \frac{1}{\rho'} d\rho', \qquad N(\rho) = \int_q^\rho A\gamma(\rho')^{\gamma-2} d\rho' \qquad (4.21)$$

with $q > 0$ being a constant independent of ρ. We write the last two terms on the left hand side of the momentum equation in (4.20) and its right hand side, into the multiplication of ρ and a term of divergence form:

- The two terms involving ρ^{n+1}.

$$\frac{1}{\varepsilon^2} a^n \nabla \rho^{n+1} - \frac{a^n}{\varepsilon^2} \rho^{n+1} \nabla \ln \rho_0 = \frac{a^n}{\varepsilon^2} \left[\nabla \rho^{n+1} - \rho^{n+1} \nabla \ln \rho_0 \right]$$

$$= \frac{a^n}{\varepsilon^2} \rho^{n+1} \left[\nabla M(\rho^{n+1}) - \nabla \ln \rho_0 \right].$$

The last equality holds due to the definition of $M(\rho)$ in (4.21).

- The two terms involving ρ^n.

$$\frac{1}{\varepsilon^2} \nabla [p(\rho) - a(t)\rho]^n + \frac{1}{\varepsilon^2} \rho^n \nabla \phi + \frac{a^n}{\varepsilon^2} \rho^n \nabla \ln \rho_0$$

$$= \frac{1}{\varepsilon^2} \left[\gamma A (\rho^n)^{\gamma-1} \nabla \rho^n - a^n \nabla \rho^n + a^n \rho^n \nabla \ln \rho_0 + \rho^n \nabla \phi \right],$$

$$= \frac{1}{\varepsilon^2} \rho^n \left[\gamma A (\rho^n)^{\gamma-2} \nabla \rho^n + \nabla \phi - a^n \frac{\nabla \rho^n}{\rho^n} + a^n \nabla \ln \rho_0 \right],$$

$$= \frac{1}{\varepsilon^2} \rho^n \left[\nabla N(\rho^n) + \nabla \phi - a^n [\nabla M(\rho^n) - \nabla \ln \rho_0] \right].$$

The last equality holds due to the definition of $N(\rho)$ in (4.21).

Hence, the momentum equation can be rewritten as,

$$\frac{(\rho \mathbf{u})^{n+1} - (\rho \mathbf{u})^n}{\Delta t} + \nabla \cdot (\rho \mathbf{u} \otimes \mathbf{u})^n + \frac{1}{\varepsilon^2} \rho^n \left[\nabla N(\rho^n) + \nabla \phi \right.$$

$$\left. - a^n [\nabla M(\rho^n) - \nabla \ln \rho_0] \right] + \frac{a^n}{\varepsilon^2} \rho^{n+1} \left[\nabla M(\rho^{n+1}) - \nabla \ln \rho_0 \right] = 0. \quad (4.22)$$

And the semi-discrete system of equations (4.20) can be rewritten as,

$$\frac{\rho^{n+1} - \rho^n}{\Delta t} + \alpha \nabla \cdot (\rho \mathbf{u})^n + (1 - \alpha) \nabla \cdot (\rho \mathbf{u})^{n+1} = 0,$$

$$\frac{(\rho \mathbf{u})^{n+1} - (\rho \mathbf{u})^n}{\Delta t} + \nabla \cdot (\rho \mathbf{u} \otimes \mathbf{u})^n + \frac{1}{\varepsilon^2} \rho^n \left[\nabla N(\rho^n) + \nabla \phi - a^n [\nabla M(\rho^n) - \nabla \ln \rho_0] \right]$$

$$+ \frac{a^n}{\varepsilon^2} \rho^{n+1} \left[\nabla M(\rho^{n+1}) - \nabla \ln \rho_0 \right] = 0.$$

$$(4.23)$$

Definition 1. (ρ, u, v) *are said to be well-prepared data if they satisfy,*

$$\rho^n = \rho^{(0)} + \varepsilon\rho^{(1)} + O(\varepsilon^2) = \left(1 - \frac{\gamma-1}{\gamma A}\phi\right)^{\frac{1}{\gamma-1}} + O(\varepsilon^2), \qquad \nabla \cdot (\rho^{(0)} u^{(0)n}) = 0.$$
(4.24)

Lemma 2. *Choose* (ρ, u, v) *to be well-prepared, then when* $\varepsilon \ll 1$,

$$\mathcal{L} = \frac{1}{\varepsilon^2}\rho^n\left[\nabla N(\rho^n) + \nabla\phi - a^n[\nabla M(\rho^n) - \nabla\ln\rho_0]\right]$$
$$= \rho^{(0)}\nabla\left[(N(\rho^n))^{(2)} - a^{(0)}(M(\rho^n))^{(2)}\right] + O(\varepsilon).$$

Proof. Let the expansion of a^n be given by

$$a^n = a^{(0)n} + \varepsilon a^{(1)n} + \varepsilon^2 a^{(2)n} + \mathcal{O}(\varepsilon^3).$$
(4.25)

Due to (4.24), from the definition $a(t) = \min_x\{p'(\rho)\}$, we find

$$a^{(0)n} = \min_x\{p'(\rho^{(0)})\} = a^{(0)}, \qquad a^{(1)n} = 0.$$
(4.26)

Moreover, the expansions of $M(\rho^n) = M^n$ and $N(\rho^n) = N^n$ around $\rho^{(0)}$ are given as

$$M^n = M(\rho^n) = M\left(\rho^{(0)} + \varepsilon\rho^{(1)} + \varepsilon^2\rho^{(2)n} + \mathcal{O}(\varepsilon^3)\right),$$
$$= M(\rho^{(0)}) + \varepsilon(\rho^{(1)} + \varepsilon\rho^{(2)n})M'(\rho^{(0)})$$
$$+ \frac{\varepsilon^2}{2}(\rho^{(1)} + \varepsilon\rho^{(2)n})^2 M''(\rho^{(0)}) + \mathcal{O}(\varepsilon^3),$$
(4.27)
$$= M(\rho^{(0)}) + \varepsilon^2\rho^{(2)n}M'(\rho^{(0)}) + \mathcal{O}(\varepsilon^3),$$
$$N^n = N(\rho^n) = N(\rho^{(0)}) + \varepsilon^2\rho^{(2)n}N'(\rho^{(0)}) + \mathcal{O}(\varepsilon^3).$$

Due to the definition of $M(\rho)$ and $\rho^{(0)}$, we have

$$\ln(\rho^{(0)})^{\gamma-1} = \ln\left(1 - \frac{\gamma-1}{\gamma A}\phi\right).$$

Thus,

$$(\gamma-1)\frac{\nabla\rho^{(0)}}{\rho^{(0)}} = \frac{-\frac{\gamma-1}{\gamma A}\nabla\phi}{(1 - \frac{\gamma-1}{\gamma A}\phi)}$$

which gives

$$(\nabla M^n)^{(0)} - \nabla\ln\rho_0 = \frac{1}{\rho^{(0)}}\nabla\rho^{(0)} - \nabla\ln\rho_0 = 0.$$

Similarly, using the definitions of $N(\rho)$ and the definition of $\rho^{(0)}$ in (4.11), we have

$$(\nabla N^n)^{(0)} + \nabla\phi = \nabla N(\rho^{(0)}) + \nabla\phi = 0.$$

Therefore, from the above results, we find

$$
\begin{aligned}
\mathcal{L} =& \frac{\rho^{(0)}}{\varepsilon^2} \Big[(\nabla N^n)^{(0)} + \nabla\phi - a^{(0)n} \big[(\nabla M^n)^{(0)} - \nabla\ln\rho_0 \big] \Big] \\
&+ \frac{\rho^{(0)}}{\varepsilon} \Big[(\nabla N^n)^{(1)} - a^{(0)n}(\nabla M^n)^{(1)} - a^{(1)n}\big[(\nabla M^n)^{(0)} - \nabla\ln\rho_0 \big] \Big] \\
&+ \frac{\rho^{(1)}}{\varepsilon} \Big[(\nabla N^n)^{(0)} + \nabla\phi - a^{(0)n}\big[(\nabla M^n)^{(0)} - \nabla\ln\rho_0 \big] \Big] \\
&+ \rho^{(0)} \Big[(\nabla N^n)^{(2)} - a^{(0)n}(\nabla M^n)^{(2)} - a^{(1)n}(\nabla M^n)^{(1)} - a^{(2)n}\big[(\nabla M^n)^{(0)} - \nabla\ln\rho_0 \big] \Big] \\
&+ \rho^{(1)} \Big[(\nabla N^n)^{(1)} - a^{(0)n}(\nabla M^n)^{(1)} - a^{(1)n}\big[(\nabla M^n)^{(0)} - \nabla\ln\rho_0 \big] \Big] \\
&+ \rho^{(2)n} \Big[(\nabla N^n)^{(0)} + \nabla\phi - a^{(0)n}\big[(\nabla M^n)^{(0)} - \nabla\ln\rho_0 \big] \Big] + \mathcal{O}(\varepsilon) \\
=& \rho^{(0)} \big[(\nabla N^n)^{(2)} - a^{(0)}(\nabla M^n)^{(2)} \big] + \mathcal{O}(\varepsilon).
\end{aligned}
$$

Hence, we conclude the proof of the lemma. □

Then we compare $\mathcal{O}(\frac{1}{\varepsilon^2})$ terms in the momentum equation in (4.23) and the only term left of order $\frac{1}{\varepsilon^2}$ is,

$$
a^{(0)}\rho^{(0)n+1}\big[(\nabla M^{n+1})^{(0)} - \nabla\ln\rho_0 \big] = 0,
$$

but $a^{(0)} \neq 0$ and $\rho^{(0)n+1} \neq 0$, thus

$$
(\nabla M^{n+1})^{(0)} = \nabla\ln\rho_0. \tag{4.28}
$$

Then

$$
\nabla ln(\rho^{(0)n+1}) = \nabla ln(1 - \frac{\gamma-1}{\gamma A}\phi)^{\frac{1}{\gamma-1}},
$$

which yields

$$
ln(\rho^{(0)n+1})^{\gamma-1} = ln(1 - \frac{\gamma-1}{\gamma A}\phi) + c,
$$

and $\rho^{(0)n+1}$ satisfies

$$
\rho^{(0)n+1} = c\left(1 - \frac{\gamma-1}{\gamma A}\phi\right)^{\frac{1}{\gamma-1}}.
$$

Here c is an arbitrary constant determined by the boundary condition of ρ. If the boundary condition of ρ does not change with time, from the definition of $\phi(x)$, we find $\rho^{(0)n+1} = \rho^{(0)}$. The $\mathcal{O}(\frac{1}{\varepsilon})$ terms in the momentum equation (4.22) are

$$
a^{(0)n}\rho^{(0)n+1}(\nabla M^{n+1})^{(1)} + \big(a^{(0)n}\rho^{(1)n+1} + a^{(1)n}\rho^{(0)n+1} \big)\nabla\big[M^{(0)n+1} - \ln\rho_0 \big] = 0.
$$

Due to (4.28), $(\nabla M^{n+1})^{(1)} = 0$. The boundary condition of ρ^{n+1} leads to $\rho^{(1)n+1} = 0$. Now compare $\mathcal{O}(1)$ terms in the density equation in (4.20),

$$
\frac{\rho^{(0)n+1} - \rho^{(0)n}}{\Delta t} + \alpha\nabla\cdot(\rho^{(0)}\mathbf{u}^{(0)})^n + (1-\alpha)\nabla\cdot(\rho^{(0)}\mathbf{u}^{(0)})^{n+1} = 0. \tag{4.29}
$$

Because $\rho^{(0)n+1} = \rho^{(0)n} = \rho^{(0)}$ is time independent and the initial data are well prepared, then equation (4.29) gives

$$\nabla \cdot (\rho^{(0)} \mathbf{u}^{(0)n+1}) = 0. \tag{4.30}$$

Compare $\mathcal{O}(1)$ terms in the momentum equation,

$$\frac{(\rho^{(0)} \mathbf{u}^{(0)})^{n+1} - (\rho^{(0)} \mathbf{u}^{(0)})^n}{\Delta t} + \nabla \cdot (\rho^{(0)} \mathbf{u}^{(0)} \otimes \mathbf{u}^{(0)})^n + \rho^{(0)} \nabla ((N^n)^{(2)} - a^{(0)} (M^n)^{(2)})$$
$$+ a^{(0)} \rho^{(0)n+1} (\nabla M^{n+1})^{(2)} + (a^{(0)} \rho^{(1)n+1} + a^{(1),n} \rho^{(0)n+1})(\nabla M^{n+1})^{(1)}$$
$$+ \left(a^{(0)} \rho^{(2)n+1} + a^{(1)n} \rho^{(1)n+1} + a^{(2)n} \rho^{(0)n+1}\right) \nabla [M^{(0)n+1} - \ln \rho_0] = 0.$$

Using the fact that $\rho^{(0)n+1} = \rho^{(0)n} = \rho^{(0)}$ is constant in time and $(\nabla M^{n+1})^{(0)} = \nabla \ln \rho_0$ and $(\nabla M^{n+1})^{(1)} = 0$, the equation simplifies to,

$$\frac{\mathbf{u}^{(0)n+1} - \mathbf{u}^{(0)n}}{\Delta t} + \mathbf{u}^{(0)n} \nabla \cdot \mathbf{u}^{(0)n} + \nabla (N^n - a^{(0)} M^n + a^{(0)} M^{n+1})^{(2)} = 0.$$

Therefore, as ε goes to zero, the solution of (4.20) converges to

$$\nabla \cdot (\rho^{(0)} \mathbf{u}^{(0)})^{n+1} = 0,$$
$$\frac{\mathbf{u}^{(0)n+1} - \mathbf{u}^{(0)n}}{\Delta t} + \mathbf{u}^{(0)n} \nabla \cdot \mathbf{u}^{(0)n} + (\nabla W^{n+1})^{(2)} = 0, \tag{4.31}$$

with $W^{n+1} = N^n - a^{(0)} M^n + a^{(0)} M^{n+1}$. Therefore, (4.31) is a good discretization of the incompressible limit equations (4.13) and the semi-discrete scheme (4.20) is AP.

4.4 Fully discrete Numerical Scheme

In order to complete the presentation of the numerical scheme, we still need space discretization. In this work, we follow the staggered discretization on a Cartesian grid suggested by Goudon et al. [34] which follows the principles of MAC schemes [38]. System (4.20) splits into two systems, the slow explicit system:

$$\begin{cases} \frac{\rho^* - \rho^n}{\Delta t} + \alpha \nabla \cdot (\rho \mathbf{u})^n = 0, \\ \frac{(\rho \mathbf{u})^* - (\rho \mathbf{u})^n}{\Delta t} + \nabla \cdot (\rho \mathbf{u} \otimes \mathbf{u})^n + \frac{1}{\varepsilon^2} \nabla [p(\rho) - a(t)\rho]^n = -\frac{1}{\varepsilon^2} \rho^n \nabla \phi - \frac{a^n}{\varepsilon^2} \rho^n \nabla \ln \rho_0, \end{cases} \tag{4.32}$$

and

$$\begin{cases} \frac{\rho^{n+1} - \rho^*}{\Delta t} + (1 - \alpha) \nabla \cdot (\rho \mathbf{u})^{n+1} = 0, \\ \frac{(\rho \mathbf{u})^{n+1} - (\rho \mathbf{u})^*}{\Delta t} + \frac{1}{\varepsilon^2} a(t^n) \nabla \rho^{n+1} = \frac{a^n}{\varepsilon^2} \rho^{n+1} \nabla \ln \rho_0. \end{cases} \tag{4.33}$$

We will deal with each system separately in one and two space dimensions.

4.4.1 The 1D numerical scheme

In the 1D setup, our computational domain $\Omega = [x_L, x_R]$, an interval of the real axis, is partitioned into subintervals $[x_i, x_{i+1}]$, for $i \in \{1, ..., N\}$. We define $x_{i+\frac{1}{2}} = \frac{x_i + x_{i+1}}{2}$ as centers of the subintervals. Let $\Delta x_i, \Delta x_{i+\frac{1}{2}}$ be the length of the interval $[x_{i-\frac{1}{2}}, x_{i+\frac{1}{2}}]$ and $[x_i, x_{i+1}]$ respectively. In our calculations, we set $\Delta x_i = \Delta x_{i+\frac{1}{2}} = \Delta x$.

The density ρ is evolved on the centers $x_{i+\frac{1}{2}}$ of the primal cells. The velocity u is evaluated on the points x_i. The density on the edges of the primal mesh can be defined by averages,

$$\rho_i = \frac{\rho_{i+\frac{1}{2}} + \rho_{i-\frac{1}{2}}}{2}. \tag{4.34}$$

We start by presenting a discretization for the slow explicit system (4.32),

$$\begin{cases} \frac{\rho^*_{i+\frac{1}{2}} - \rho^n_{i+\frac{1}{2}}}{\Delta t} + \alpha \left[\frac{F_{i+1} - F_i}{\Delta x} \right] = 0, \\ \\ \frac{\rho^*_i u^*_i - \rho^n_i u^n_i}{\Delta t} + \frac{\zeta_{i+\frac{1}{2}} - \zeta_{i-\frac{1}{2}}}{\Delta x} + \frac{1}{\varepsilon^2} \frac{\Pi^n_{i+\frac{1}{2}} - \Pi^n_{i-\frac{1}{2}}}{\Delta x} = -\frac{1}{\varepsilon^2} \rho^n_i \frac{\phi_{i+\frac{1}{2}} - \phi_{i-\frac{1}{2}}}{\Delta x} \\ \\ -\frac{a^n}{\varepsilon^2} \rho^n_i \frac{\ln \rho_{0,i+\frac{1}{2}} - \ln \rho_{0,i-\frac{1}{2}}}{\Delta x}. \end{cases} \tag{4.35}$$

With $\Pi^n_{i+\frac{1}{2}}$ is the modified pressure term at the node $x_{i+\frac{1}{2}}$ and is defined as

$$\Pi^n_{i+\frac{1}{2}} = p(\rho^n_{i+\frac{1}{2}}) - a^n_d \rho^n_{i+\frac{1}{2}}.$$

a^n_d is the discrete version of the time dependent function $a(t)$ at time t^n, defined as

$$a^n_d = \min_i \left\{ p'(\rho^n_{i+\frac{1}{2}}) \right\} - l\varepsilon^2.$$

The flux terms in the density equation are computed with the following formula,

$$F_i = F^+_i + F^-_i = F^+(\rho_{i-\frac{1}{2}}, u_i) + F^-(\rho_{i+\frac{1}{2}}, u_i),$$

with

$$F^+(\rho, u) = \begin{cases} 0 & \text{if} \quad u \le -c(\rho, u) \\ \frac{\rho}{4c(\rho, u)}(v + c(\rho, u))^2 & \text{if} \quad |u| \le c(\rho, u) \\ \rho u & \text{if} \quad u \ge c(\rho, u) \end{cases}$$

$$F^-(\rho, u) = \begin{cases} \rho u & \text{if} \quad u \le -c(\rho, u) \\ -\frac{\rho}{4c(\rho, u)}(v - c(\rho, u))^2 & \text{if} \quad |u| \le c(\rho, u) \\ 0 & \text{if} \quad u \ge c(\rho, u) \end{cases}$$

The flux terms in the momentum equation are computed as the following,

$$\zeta_{i+\frac{1}{2}} = u_i F^+_{i+\frac{1}{2}} + u_{i+1} F^+_{i+\frac{1}{2}},$$

with

$$F^{\pm}_{i+\frac{1}{2}} = \frac{1}{2}\left(F^{\pm}_i + F^{\pm}_{i+1}\right).$$

The next step is to discretize the fast implicit system (4.33),

$$\begin{cases} \frac{\rho^{n+1}_{i+\frac{1}{2}} - \rho^*_{i+\frac{1}{2}}}{\Delta t} + (1-\alpha)\left[\frac{(F^{n+1})^{Up}_{i+1} - (F^{n+1})^{Up}_i}{\Delta x}\right] = 0, \\[3mm] \frac{\rho^{n+1}_i u^{n+1}_i - \rho^*_i u^*_i}{\Delta t} + \frac{a^n_d}{\varepsilon^2}\frac{\rho^{n+1}_{i+\frac{1}{2}} - \rho^{n+1}_{i-\frac{1}{2}}}{\Delta x} = \frac{a^n_d}{\varepsilon^2}\rho^{n+1}_i \frac{\ln\rho_{0,i+\frac{1}{2}} - \ln\rho_{0,i-\frac{1}{2}}}{\Delta x}. \end{cases} \tag{4.36}$$

Here $(F^{n+1})^{Up}_i$ is the upwind flux function obtained as following,

$$(F^{n+1})^{Up}_i = \rho^{n+1}_{i-\frac{1}{2}}\left[u^{n+1}_i\right]^+ - \rho^{n+1}_{i+\frac{1}{2}}\left[u^{n+1}_i\right]^-.$$

Where $[X]^+ = \frac{|X|+X}{2}$. Now, in order to solve this implicit system we first write u^{n+1}_i from the momentum equation in (4.36) as a function of ρ^{n+1},

$$u^{n+1}_i = \frac{1}{\rho^{n+1}_i}\left[\rho^*_i u^*_i - \frac{a^n\Delta t}{\varepsilon^2}\frac{\rho^{n+1}_{i+\frac{1}{2}} - \rho_{i-\frac{1}{2}}}{\Delta x} + \frac{a^n\Delta t}{\varepsilon^2}\rho^{n+1}_i\frac{\ln\rho_{0,i+\frac{1}{2}} - \ln\rho_{0,i-\frac{1}{2}}}{\Delta x}\right]. \tag{4.37}$$

Substitute the flux terms by their values in the density equation,

$$\frac{\rho^{n+1}_{i+\frac{1}{2}} - \rho^*_{i+\frac{1}{2}}}{\Delta t} + \frac{(1-\alpha)}{\Delta x}\left[\rho^{n+1}_{i+\frac{1}{2}}\left[u^{n+1}_{i+1}\right]^+ - \rho^{n+1}_{i+\frac{3}{2}}\left[u^{n+1}_{i+1}\right]^- \right.$$
$$\left. - \rho^{n+1}_{i-\frac{1}{2}}\left[u^{n+1}_i\right]^+ + \rho^{n+1}_{i+\frac{1}{2}}\left[u^{n+1}_i\right]^-\right] = 0. \tag{4.38}$$

Keeping in mind that $[X]^+ = \frac{|X|+X}{2}$,

$$\frac{\rho^{n+1}_{i+\frac{1}{2}} - \rho^*_{i+\frac{1}{2}}}{\Delta t} + \frac{(1-\alpha)}{\Delta x}\left[\rho^{n+1}_{i+\frac{1}{2}}\frac{|u^{n+1}_{i+1}| + u^{n+1}_{i+1}}{2} - \rho^{n+1}_{i+\frac{3}{2}}\frac{|u^{n+1}_{i+1}| - u^{n+1}_{i+1}}{2}\right.$$
$$\left. - \rho^{n+1}_{i-\frac{1}{2}}\frac{|u^{n+1}_i| + u^{n+1}_i}{2} + \rho^{n+1}_{i+\frac{1}{2}}\frac{|u^{n+1}_i| - u^{n+1}_i}{2}\right] = 0. \tag{4.39}$$

Rearranging the terms yields to,

$$\frac{\rho^{n+1}_{i+\frac{1}{2}} - \rho^*_{i+\frac{1}{2}}}{\Delta t} + \frac{(1-\alpha)}{\Delta x}\left[\frac{\rho^{n+1}_{i+\frac{1}{2}} - \rho^{n+1}_{i+\frac{3}{2}}}{2}|u^{n+1}_{i+1}| + \frac{\rho^{n+1}_{i+\frac{1}{2}} + \rho^{n+1}_{i+\frac{3}{2}}}{2}u^{n+1}_{i+1}\right.$$
$$\left. + \frac{\rho^{n+1}_{i+\frac{1}{2}} - \rho^{n+1}_{i-\frac{1}{2}}}{2}|u^{n+1}_i| - \frac{\rho^{n+1}_{i+\frac{1}{2}} + \rho^{n+1}_{i-\frac{1}{2}}}{2}u^{n+1}_i\right] = 0. \tag{4.40}$$

However, from the definition of ρ_i^{n+1}, the equation can be rewritten into this

$$
\frac{\rho_{i+\frac{1}{2}}^{n+1} - \rho_{i+\frac{1}{2}}^*}{\Delta t} + \frac{(1-\alpha)}{\Delta x} \left[\frac{\rho_{i+\frac{1}{2}}^{n+1} - \rho_{i+\frac{3}{2}}^{n+1}}{2} |u_{i+1}^{n+1}| + \rho_{i+1}^{n+1} u_{i+1}^{n+1} \right.
$$
$$
\left. + \frac{\rho_{i+\frac{1}{2}}^{n+1} - \rho_{i-\frac{1}{2}}^{n+1}}{2} |u_i^{n+1}| - \rho_i^{n+1} u_i^{n+1} \right] = 0. \quad (4.41)
$$

Next, we substitute u_i^{n+1} by its value in the density equation,

$$
\frac{\rho_{i+\frac{1}{2}}^{n+1} - \rho_{i+\frac{1}{2}}^*}{\Delta t} + \frac{(1-\alpha)}{\Delta x} \left[\frac{\rho_{i+\frac{1}{2}}^{n+1} - \rho_{i+\frac{3}{2}}^{n+1}}{2\rho_{i+1}^{n+1}} \right.
$$
$$
\left| \rho_{i+1}^* u_{i+1}^* - \frac{a^n \Delta t}{\varepsilon^2} \frac{\rho_{i+\frac{3}{2}}^{n+1} - \rho_{i+\frac{1}{2}}}{\Delta x} + \frac{a^n \Delta t}{\varepsilon^2} \rho_{i+1}^{n+1} \frac{\ln \rho_{0,i+\frac{3}{2}} - \ln \rho_{0,i+\frac{1}{2}}}{\Delta x} \right|
$$
$$
+ \left(\rho_{i+1}^* u_{i+1}^* - \frac{a^n \Delta t}{\varepsilon^2} \frac{\rho_{i+\frac{3}{2}}^{n+1} - \rho_{i+\frac{1}{2}}}{\Delta x} + \frac{a^n \Delta t}{\varepsilon^2} \rho_{i+1}^{n+1} \frac{\ln \rho_{0,i+\frac{3}{2}} - \ln \rho_{0,i+\frac{1}{2}}}{\Delta x} \right)
$$
$$
+ \frac{\rho_{i+\frac{1}{2}}^{n+1} - \rho_{i-\frac{1}{2}}^{n+1}}{2\rho_i^{n+1}}
$$
$$
\left| \rho_i^* u_i^* - \frac{a^n \Delta t}{\varepsilon^2} \frac{\rho_{i+\frac{1}{2}}^{n+1} - \rho_{i-\frac{1}{2}}}{\Delta x} + \frac{a^n \Delta t}{\varepsilon^2} \rho_i^{n+1} \frac{\ln \rho_{0,i+\frac{1}{2}} - \ln \rho_{0,i-\frac{1}{2}}}{\Delta x} \right|
$$
$$
\left. - \left(\rho_i^* u_i^* - \frac{a^n \Delta t}{\varepsilon^2} \frac{\rho_{i+\frac{1}{2}}^{n+1} - \rho_{i-\frac{1}{2}}}{\Delta x} + \frac{a^n \Delta t}{\varepsilon^2} \rho_i^{n+1} \frac{\ln \rho_{0,i+\frac{1}{2}} - \ln \rho_{0,i-\frac{1}{2}}}{\Delta x} \right) \right] = 0. \quad (4.42)
$$

The previous system of N nonlinear equations is to be solved using the Newton-Raphson method. We are interested in solving the system $\mathbf{f(x)=0}$ with
$\mathbf{0} = [0.....0]^T$,
$\mathbf{x} = [x_1 x_2 ... x_N]^T = \left[\rho_{\frac{3}{2}}^{n+1} \rho_{\frac{5}{2}}^{n+1} ... \rho_{N+\frac{1}{2}}^{n+1} \right]^T$,
$\mathbf{f(x)} = [f_1(\mathbf{x}) f_2(\mathbf{x}) ... f_N(\mathbf{x})]^T$,

where T denotes the transpose operator and $f_i(\mathbf{x}) =$

$$
\frac{\rho_{i+\frac{1}{2}}^{n+1} - \rho_{i+\frac{1}{2}}^{*}}{\Delta t} + \frac{(1-\alpha)}{\Delta x} \left[\frac{\rho_{i+\frac{1}{2}}^{n+1} - \rho_{i+\frac{3}{2}}^{n+1}}{2\rho_{i+1}^{n+1}} \right.
$$

$$
\left| \rho_{i+1}^{*} u_{i+1}^{*} - \frac{a^n \Delta t}{\varepsilon^2} \frac{\rho_{i+\frac{3}{2}}^{n+1} - \rho_{i+\frac{1}{2}}}{\Delta x} + \frac{a^n \Delta t}{\varepsilon^2} \rho_{i+1}^{n+1} \frac{\ln \rho_{0,i+\frac{3}{2}} - \ln \rho_{0,i+\frac{1}{2}}}{\Delta x} \right|
$$

$$
+ \left(\rho_{i+1}^{*} u_{i+1}^{*} - \frac{a^n \Delta t}{\varepsilon^2} \frac{\rho_{i+\frac{3}{2}}^{n+1} - \rho_{i+\frac{1}{2}}}{\Delta x} + \frac{a^n \Delta t}{\varepsilon^2} \rho_{i+1}^{n+1} \frac{\ln \rho_{0,i+\frac{3}{2}} - \ln \rho_{0,i+\frac{1}{2}}}{\Delta x} \right)
$$

$$
+ \frac{\rho_{i+\frac{1}{2}}^{n+1} - \rho_{i-\frac{1}{2}}^{n+1}}{2\rho_i^{n+1}}
$$

$$
\left| \rho_i^{*} u_i^{*} - \frac{a^n \Delta t}{\varepsilon^2} \frac{\rho_{i+\frac{1}{2}}^{n+1} - \rho_{i-\frac{1}{2}}}{\Delta x} + \frac{a^n \Delta t}{\varepsilon^2} \rho_i^{n+1} \frac{\ln \rho_{0,i+\frac{1}{2}} - \ln \rho_{0,i-\frac{1}{2}}}{\Delta x} \right|
$$

$$
\left. - \left(\rho_i^{*} u_i^{*} - \frac{a^n \Delta t}{\varepsilon^2} \frac{\rho_{i+\frac{1}{2}}^{n+1} - \rho_{i-\frac{1}{2}}}{\Delta x} + \frac{a^n \Delta t}{\varepsilon^2} \rho_i^{n+1} \frac{\ln \rho_{0,i+\frac{1}{2}} - \ln \rho_{0,i-\frac{1}{2}}}{\Delta x} \right) \right]. \quad (4.43)
$$

After solving the system of nonlinear equations for ρ^{n+1}, $(\rho u)^{n+1}$ is recovered from the momentum equation. The full presentation of the 1D scheme is summarized by the slow step (4.35) together with the fast step (4.36).

4.4.2 The 2D numerical scheme

The space discretization follows the idea of a recent work by Goudon et.al. [34], where the authors design an AP scheme for the isentropic Euler equations. We consider a computational domain $[x_L, x_R] \times [y_L, y_R]$ and Cartesian 2D grid points. The grid points are x_i, y_j for $i, j \in \{1, ..., N_x\}$ and $j \in \{1, ..., N_y\}$ and we define $x_{i+\frac{1}{2}} = \frac{x_i + x_{i+1}}{2}$ and $y_{j+\frac{1}{2}} = \frac{y_j + y_{j+1}}{2}$ for $i \in \{1, ..., N_x - 1\}$, $j \in \{1, ..., N_y - 1\}$. Let $\Delta x_i, \Delta x_{i+\frac{1}{2}}, \Delta y_j$, and $\Delta y_{j+\frac{1}{2}}$ be respectively the length of the interval $[x_{i-\frac{1}{2}}, x_{i+\frac{1}{2}}]$, $[x_i, x_{i+1}], [y_{j-\frac{1}{2}}, y_{j+\frac{1}{2}}]$ and $[y_j, y_{j+1}]$. In our calculations, we set $\Delta x_i = \Delta x_{i+\frac{1}{2}} = \Delta x$ and $\Delta y_j = \Delta y_{j+\frac{1}{2}} = \Delta y$. Let Δt be the time step. As in Figure 4.1, the density ρ is defined at the points $(x_{i+\frac{1}{2}}, y_{j+\frac{1}{2}})$, while the velocity u in the x-direction is evaluated on the points $(x_i, y_{j+\frac{1}{2}})$ and the velocity v in the y-direction is evaluated on the points $(x_{i+\frac{1}{2}}, y_j)$. The density on the edges of the primal mesh can be defined by the average value of two neighbouring cells such that

$$
\rho_{i,j+\frac{1}{2}} = \frac{\rho_{i+\frac{1}{2},j+\frac{1}{2}} + \rho_{i-\frac{1}{2},j+\frac{1}{2}}}{2},
$$

$$
\rho_{i+\frac{1}{2},j} = \frac{\rho_{i+\frac{1}{2},j+\frac{1}{2}} + \rho_{i+\frac{1}{2},j-\frac{1}{2}}}{2}.
$$

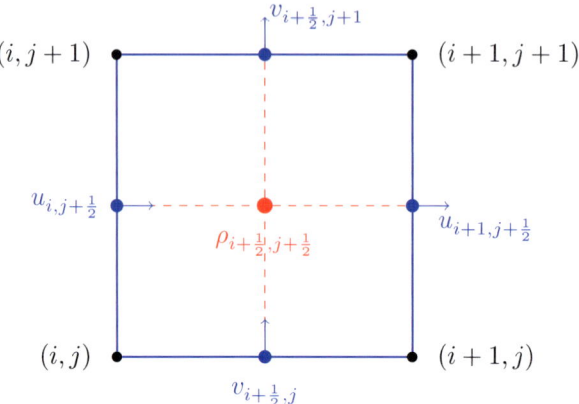

Figure 4.1: MAC discretization.

The numerical solution is evolved on the staggered grid and the fluxes are defined as in [34]. We start by presenting a discretization for the slow explicit system (4.32),

$$
\begin{cases}
\dfrac{\rho^*_{i+\frac{1}{2},j+\frac{1}{2}}-\rho^n_{i+\frac{1}{2},j+\frac{1}{2}}}{\Delta t} + \alpha\left[\dfrac{F^{x,n}_{i+1,j+\frac{1}{2}}-F^{x,n}_{i,j+\frac{1}{2}}}{\Delta x} + \dfrac{F^{y,n}_{i+\frac{1}{2},j+1}-F^{y,n}_{i+\frac{1}{2},j}}{\Delta y}\right] = 0,\\[4mm]
\dfrac{\rho^*_{i,j+\frac{1}{2}}u^*_{i,j+\frac{1}{2}}-\rho^n_{i,j+\frac{1}{2}}u^n_{i,j+\frac{1}{2}}}{\Delta t} + \dfrac{\zeta^{u,x}_{i+\frac{1}{2},j+\frac{1}{2}}-\zeta^{u,x}_{i-\frac{1}{2},j+\frac{1}{2}}}{\Delta x} + \dfrac{\zeta^{u,y}_{i,j+1}-\zeta^{u,y}_{i,j}}{\Delta y} + \dfrac{1}{\varepsilon^2}\dfrac{\Pi^n_{i+\frac{1}{2},j+\frac{1}{2}}-\Pi^n_{i-\frac{1}{2},j+\frac{1}{2}}}{\Delta x}\\[3mm]
\quad = -\dfrac{1}{\varepsilon^2}\rho^n_{i,j+\frac{1}{2}}\dfrac{\phi_{i+\frac{1}{2},j+\frac{1}{2}}-\phi_{i-\frac{1}{2},j+\frac{1}{2}}}{\Delta x} - \dfrac{a^n}{\varepsilon^2}\rho^n_{i,j+\frac{1}{2}}\dfrac{\ln\rho_{0,i+\frac{1}{2},j+\frac{1}{2}}-\ln\rho_{0,i-\frac{1}{2},j+\frac{1}{2}}}{\Delta x},\\[4mm]
\dfrac{\rho^*_{i+\frac{1}{2},j}v^*_{i+\frac{1}{2},j}-\rho^n_{i+\frac{1}{2},j}v^n_{i+\frac{1}{2},j}}{\Delta t} + \dfrac{\zeta^{v,x}_{i+1,j}-\zeta^{v,x}_{i,j}}{\Delta x} + \dfrac{\zeta^{v,y}_{i+\frac{1}{2},j+\frac{1}{2}}-\zeta^{v,y}_{i+\frac{1}{2},j-\frac{1}{2}}}{\Delta y} + \dfrac{1}{\varepsilon^2}\dfrac{\Pi^n_{i+\frac{1}{2},j+\frac{1}{2}}-\Pi^n_{i+\frac{1}{2},j-\frac{1}{2}}}{\Delta y}\\[3mm]
\quad = -\dfrac{1}{\varepsilon^2}\rho^n_{i+\frac{1}{2},j}\dfrac{\phi_{i+\frac{1}{2},j+\frac{1}{2}}-\phi_{i+\frac{1}{2},j-\frac{1}{2}}}{\Delta y} - \dfrac{a^n}{\varepsilon^2}\rho^n_{i+\frac{1}{2},j}\dfrac{\ln\rho_{0,i+\frac{1}{2},j+\frac{1}{2}}-\ln\rho_{0,i+\frac{1}{2},j-\frac{1}{2}}}{\Delta y},
\end{cases}
$$

$$(4.44)$$

with

$$
\Pi^n_{i+\frac{1}{2},j+\frac{1}{2}} = P\big(\rho^n_{i+\frac{1}{2},j+\frac{1}{2}}\big) - a^n\rho^n_{i+\frac{1}{2},j+\frac{1}{2}}
$$

where a^n is the discrete version of $a(t)$ defined as,

$$
a^n = \min_{i,j}\left\{P'\big(\rho^n_{i+\frac{1}{2},j+\frac{1}{2}}\big)\right\} - l\varepsilon^2.
$$

The flux terms are

$$
F^x_{i,j+\frac{1}{2}} - F^{x,+}_{i,j+\frac{1}{2}} + F^{x,-}_{i,j+\frac{1}{2}} = F^+\big(\rho_{i-\frac{1}{2},j+\frac{1}{2}},u_{i,j+\frac{1}{2}}\big) + F^-\big(\rho_{i+\frac{1}{2},j+\frac{1}{2}},u_{i,j+\frac{1}{2}}\big),
$$

$$
F^y_{i+\frac{1}{2},j} = F^{y,+}_{i+\frac{1}{2},j} + F^{y,-}_{i+\frac{1}{2},j} = F^+\big(\rho_{i+\frac{1}{2},j-\frac{1}{2}},v_{i+\frac{1}{2},j}\big) + F^-\big(\rho_{i+\frac{1}{2},j+\frac{1}{2}},v_{i+\frac{1}{2},j}\big),
$$

$$\zeta^{u,x}_{i+\frac{1}{2},j+\frac{1}{2}} = u_{i,j+\frac{1}{2}} F^{x,+}_{i+\frac{1}{2},j+\frac{1}{2}} + u_{i+1,j+\frac{1}{2}} F^{x,+}_{i+\frac{1}{2},j+\frac{1}{2}}, \zeta^{u,y}_{i,j} = u_{i,j-\frac{1}{2}} F^{y,+}_{i,j} + u_{i,j+\frac{1}{2}} F^{y,+}_{i,j}$$

$$\zeta^{v,x}_{i,j} = v_{i-\frac{1}{2},j} F^{x,+}_{i,j} + v_{i+\frac{1}{2},j} F^{x,+}_{i,j}, \zeta^{v,y}_{i+\frac{1}{2},j+\frac{1}{2}} = v_{i+\frac{1}{2},j} F^{y,+}_{i+\frac{1}{2},j+\frac{1}{2}} + v_{i+\frac{1}{2},j+1} F^{y,+}_{i+\frac{1}{2},j+\frac{1}{2}}$$

with

$$F^{+}(\rho,u) = \begin{cases} 0, & \text{if} \quad u \leq -c(\rho,u), \\ \frac{\rho}{4c(\rho,u)}(v + c(\rho,u))^2, & \text{if} \quad |u| \leq c(\rho,u), \\ \rho u, & \text{if} \quad u \geq c(\rho,u), \end{cases}$$

$$F^{-}(\rho,u) = \begin{cases} \rho u, & \text{if} \quad u \leq -c(\rho,u), \\ -\frac{\rho}{4c(\rho,u)}(v - c(\rho,u))^2, & \text{if} \quad |u| \leq c(\rho,u), \\ 0, & \text{if} \quad u \geq c(\rho,u), \end{cases}$$

and

$$F^{x,\pm}_{i+\frac{1}{2},j+\frac{1}{2}} = \frac{1}{2}\left(F^{x,\pm}_{i,j+\frac{1}{2}} + F^{x,\pm}_{i+1,j+\frac{1}{2}}\right), \qquad F^{y,\pm}_{i,j} = \frac{1}{2}\left(F^{y,\pm}_{i+\frac{1}{2},j} + F^{y,\pm}_{i-\frac{1}{2},j}\right).$$

The flux terms in the momentum equation along the x-direction are computed as follows,

$$\zeta^{u,x}_{i+\frac{1}{2},j+\frac{1}{2}} = u_{i,j+\frac{1}{2}} F^{x,+}_{i+\frac{1}{2},j+\frac{1}{2}} + u_{i+1,j+\frac{1}{2}} F^{x,-}_{i+\frac{1}{2},j+\frac{1}{2}},$$

$$= u_{i,j+\frac{1}{2}} \frac{F^{x,+}_{i,j+\frac{1}{2}} + F^{x,+}_{i+1,j+\frac{1}{2}}}{2} + u_{i+1,j+\frac{1}{2}} \frac{F^{x,-}_{i,j+\frac{1}{2}} + F^{x,-}_{i+1,j+\frac{1}{2}}}{2},$$

$$= u_{i,j+\frac{1}{2}} \frac{F^{+}(\rho_{i-\frac{1}{2},j+\frac{1}{2}}, u_{i,j+\frac{1}{2}}) + F^{+}(\rho_{i+\frac{1}{2},j+\frac{1}{2}}, u_{i+1,j+\frac{1}{2}})}{2}$$

$$+ u_{i+1,j+\frac{1}{2}} \frac{F^{-}(\rho_{i+\frac{1}{2},j+\frac{1}{2}}, u_{i,j+\frac{1}{2}}) + F^{-}(\rho_{i+\frac{3}{2},j+\frac{1}{2}}, u_{i+1,j+\frac{1}{2}})}{2}.$$

$$\zeta^{u,x}_{i-\frac{1}{2},j+\frac{1}{2}} = u_{i-1,j+\frac{1}{2}} F^{x,+}_{i-\frac{1}{2},j+\frac{1}{2}} + u_{i,j+\frac{1}{2}} F^{x,-}_{i-\frac{1}{2},j+\frac{1}{2}},$$

$$= u_{i-1,j+\frac{1}{2}} \frac{F^{x,+}_{i-1,j+\frac{1}{2}} + F^{x,+}_{i,j+\frac{1}{2}}}{2} + u_{i,j+\frac{1}{2}} \frac{F^{x,-}_{i-1,j+\frac{1}{2}} + F^{x,-}_{i,j+\frac{1}{2}}}{2},$$

$$= u_{i-1,j+\frac{1}{2}} \frac{F^{+}(\rho_{i-\frac{3}{2},j+\frac{1}{2}}, u_{i-1,j+\frac{1}{2}}) + F^{+}(\rho_{i-\frac{1}{2},j+\frac{1}{2}}, u_{i,j+\frac{1}{2}})}{2}$$

$$+ u_{i,j+\frac{1}{2}} \frac{F^{-}(\rho_{i-\frac{1}{2},j+\frac{1}{2}}, u_{i-1,j+\frac{1}{2}}) + F^{-}(\rho_{i+\frac{1}{2},j+\frac{1}{2}}, u_{i,j+\frac{1}{2}})}{2}.$$

$$\zeta_{i,j}^{u,y} = u_{i,j-\frac{1}{2}} F_{i,j}^{y,+} + u_{i,j+\frac{1}{2}} F_{i,j}^{y,-},$$

$$= u_{i,j-\frac{1}{2}} \frac{F_{i+\frac{1}{2},j}^{y,+} + F_{i-\frac{1}{2},j}^{y,+}}{2} + u_{i,j+\frac{1}{2}} \frac{F_{i+\frac{1}{2},j}^{y,-} + F_{i-\frac{1}{2},j}^{y,-}}{2},$$

$$= u_{i,j-\frac{1}{2}} \frac{F^+(\rho_{i+\frac{1}{2},j-\frac{1}{2}}, v_{i+\frac{1}{2},j}) + F^+(\rho_{i-\frac{1}{2},j-\frac{1}{2}}, v_{i-\frac{1}{2},j})}{2}$$

$$+ u_{i,j+\frac{1}{2}} \frac{F^-(\rho_{i+\frac{1}{2},j+\frac{1}{2}}, v_{i+\frac{1}{2},j}) + F^-(\rho_{i-\frac{1}{2},j+\frac{1}{2}}, v_{i-\frac{1}{2},j})}{2}.$$

$$\zeta_{i,j+1}^{u,y} = u_{i,j+\frac{1}{2}} F_{i,j+1}^{y,+} + u_{i,j+\frac{3}{2}} F_{i,j+1}^{y,-},$$

$$= u_{i,j+\frac{1}{2}} \frac{F_{i+\frac{1}{2},j+1}^{y,+} + F_{i-\frac{1}{2},j+1}^{y,+}}{2} + u_{i,j+\frac{3}{2}} \frac{F_{i+\frac{1}{2},j+1}^{y,-} + F_{i-\frac{1}{2},j+1}^{y,-}}{2},$$

$$= u_{i,j+\frac{1}{2}} \frac{F^+(\rho_{i+\frac{1}{2},j+\frac{1}{2}}, v_{i+\frac{1}{2},j+1}) + F^+(\rho_{i-\frac{1}{2},j-\frac{1}{2}}, v_{i-\frac{1}{2},j})}{2}$$

$$+ u_{i,j+\frac{3}{2}} \frac{F^-(\rho_{i+\frac{1}{2},j+\frac{3}{2}}, v_{i+\frac{1}{2},j+1}) + F^-(\rho_{i-\frac{1}{2},j+\frac{3}{2}}, v_{i-\frac{1}{2},j+1})}{2}.$$

Now, in a similar manner we compute the flux terms in the momentum equation along the y-direction,

$$\zeta_{i,j}^{v,x} = v_{i-\frac{1}{2},j} F_{i,j}^{x,+} + v_{i+\frac{1}{2},j} F_{i,j}^{x,-},$$

$$= v_{i-\frac{1}{2},j} \frac{F_{i,j+\frac{1}{2}}^{x,+} + F_{i,j-\frac{1}{2}}^{x,+}}{2} + v_{i+\frac{1}{2},j} \frac{F_{i,j+\frac{1}{2}}^{x,-} + F_{i,j-\frac{1}{2}}^{x,-}}{2},$$

$$= v_{i-\frac{1}{2},j} \frac{F^+(\rho_{i-\frac{1}{2},j+\frac{1}{2}}, u_{i,j+\frac{1}{2}}) + F^+(\rho_{i-\frac{1}{2},j-\frac{1}{2}}, u_{i,j-\frac{1}{2}})}{2}$$

$$+ v_{i+\frac{1}{2},j} \frac{F^-(\rho_{i+\frac{1}{2},j+\frac{1}{2}}, u_{i,j+\frac{1}{2}}) + F^-(\rho_{i+\frac{1}{2},j-\frac{1}{2}}, u_{i,j-\frac{1}{2}})}{2}.$$

$$\zeta_{i+1,j}^{v,x} = v_{i+\frac{1}{2},j} F_{i+1,j}^{x,+} + v_{i+\frac{3}{2},j} F_{i+1,j}^{x,-},$$

$$= v_{i+\frac{1}{2},j} \frac{F_{i+1,j+\frac{1}{2}}^{x,+} + F_{i+1,j-\frac{1}{2}}^{x,+}}{2} + v_{i+\frac{3}{2},j} \frac{F_{i+1,j+\frac{1}{2}}^{x,-} + F_{i+1,j-\frac{1}{2}}^{x,-}}{2},$$

$$= v_{i+\frac{1}{2},j} \frac{F^+(\rho_{i+\frac{1}{2},j+\frac{1}{2}}, u_{i+1,j+\frac{1}{2}}) + F^+(\rho_{i+\frac{1}{2},j-\frac{1}{2}}, u_{i+1,j-\frac{1}{2}})}{2}$$

$$+ v_{i+\frac{3}{2},j} \frac{F^-(\rho_{i+\frac{3}{2},j+\frac{1}{2}}, u_{i+1,j+\frac{1}{2}}) + F^-(\rho_{i+\frac{3}{2},j-\frac{1}{2}}, u_{i+1,j-\frac{1}{2}})}{2}.$$

$$\zeta^{v,y}_{i+\frac{1}{2},j+\frac{1}{2}} = v_{i+\frac{1}{2},j}F^{y,+}_{i+\frac{1}{2},j+\frac{1}{2}} + v_{i+\frac{1}{2},j+1}F^{y,-}_{i+\frac{1}{2},j+\frac{1}{2}},$$

$$= v_{i+\frac{1}{2},j}\frac{F^{y,+}_{i+\frac{1}{2},j} + F^{y,+}_{i+\frac{1}{2},j+1}}{2} + v_{i+\frac{1}{2},j+1}\frac{F^{y,-}_{i+\frac{1}{2},j} + F^{y,-}_{i+\frac{1}{2},j+1}}{2},$$

$$= v_{i+\frac{1}{2},j}\frac{F^{+}(\rho_{i+\frac{1}{2},j-\frac{1}{2}},v_{i+\frac{1}{2},j}) + F^{+}(\rho_{i+\frac{1}{2},j+\frac{1}{2}},v_{i+\frac{1}{2},j+1})}{2}$$

$$+ v_{i+\frac{1}{2},j+1}\frac{F^{-}(\rho_{i+\frac{1}{2},j+\frac{1}{2}},v_{i+\frac{1}{2},j}) + F^{-}(\rho_{i+\frac{1}{2},j+\frac{3}{2}},v_{i+\frac{1}{2},j+1})}{2}.$$

$$\zeta^{v,y}_{i+\frac{1}{2},j-\frac{1}{2}} = v_{i+\frac{1}{2},j-1}F^{y,+}_{i+\frac{1}{2},j-\frac{1}{2}} + v_{i+\frac{1}{2},j}F^{y,-}_{i+\frac{1}{2},j-\frac{1}{2}},$$

$$= v_{i+\frac{1}{2},j-1}\frac{F^{y,+}_{i+\frac{1}{2},j-1} + F^{y,+}_{i+\frac{1}{2},j}}{2} + v_{i+\frac{1}{2},j}\frac{F^{y,-}_{i+\frac{1}{2},j-1} + F^{y,-}_{i+\frac{1}{2},j}}{2},$$

$$= v_{i+\frac{1}{2},j-1}\frac{F^{+}(\rho_{i+\frac{1}{2},j-\frac{3}{2}},v_{i+\frac{1}{2},j-1}) + F^{+}(\rho_{i+\frac{1}{2},j-\frac{1}{2}},v_{i+\frac{1}{2},j})}{2}$$

$$+ v_{i+\frac{1}{2},j}\frac{F^{-}(\rho_{i+\frac{1}{2},j-\frac{1}{2}},v_{i+\frac{1}{2},j-1}) + F^{-}(\rho_{i+\frac{1}{2},j+\frac{1}{2}},v_{i+\frac{1}{2},j})}{2}.$$

The next step is to discretize the fast implicit system (4.33) by

$$\begin{cases} \dfrac{\rho^{n+1}_{i+\frac{1}{2},j+\frac{1}{2}} - \rho^{*}_{i+\frac{1}{2},j+\frac{1}{2}}}{\Delta t} \\ +(1-\alpha)\left[\dfrac{(F^{n+1})^{Up,x}_{i+1,j+\frac{1}{2}} - (F^{n+1})^{Up,x}_{i,j+\frac{1}{2}}}{\Delta x} + \dfrac{(F^{n+1})^{Up,y}_{i+\frac{1}{2},j+1} - (F^{n+1})^{Up,y}_{i+\frac{1}{2},j}}{\Delta y}\right] = 0, \\ \dfrac{\rho^{n+1}_{i,j+\frac{1}{2}}u^{n+1}_{i,j+\frac{1}{2}} - \rho^{*}_{i,j+\frac{1}{2}}u^{*}_{i,j+\frac{1}{2}}}{\Delta t} \\ +\dfrac{a^{n}_{d}}{\varepsilon^2}\dfrac{\rho^{n+1}_{i+\frac{1}{2},j+\frac{1}{2}} - \rho^{n+1}_{i-\frac{1}{2},j+\frac{1}{2}}}{\Delta x} = \dfrac{a^{n}_{d}}{\varepsilon^2}\rho^{n+1}_{i,j+\frac{1}{2}}\dfrac{\ln\rho_{0,i+\frac{1}{2},j+\frac{1}{2}} - \ln\rho_{0,i-\frac{1}{2},j+\frac{1}{2}}}{\Delta x}, \\ \dfrac{\rho^{n+1}_{i+\frac{1}{2},j}v^{n+1}_{i+\frac{1}{2},j} - \rho^{*}_{i+\frac{1}{2},j}v^{*}_{i+\frac{1}{2},j}}{\Delta t} \\ +\dfrac{a^{n}}{\varepsilon^2}\dfrac{\rho^{n+1}_{i+\frac{1}{2},j+\frac{1}{2}} - \rho^{n+1}_{i+\frac{1}{2},j-\frac{1}{2}}}{\Delta y} = \dfrac{a^{n}}{\varepsilon^2}\rho^{n+1}_{i+\frac{1}{2},j}\dfrac{\ln\rho_{0,i+\frac{1}{2},j+\frac{1}{2}} - \ln\rho_{0,i+\frac{1}{2},j-\frac{1}{2}}}{\Delta y}, \end{cases} \quad (4.45)$$

where $(F^{n+1})^{Up,x}_{i,j+\frac{1}{2}}$ and $(F^{n+1})^{Up,y}_{i+\frac{1}{2},j}$ are the upwind fluxes defined as following,

$$(F^{n+1})^{Up,x}_{i,j+\frac{1}{2}} = \rho^{n+1}_{i-\frac{1}{2},j+\frac{1}{2}}\left[u^{n+1}_{i,j+\frac{1}{2}}\right]^{+} - \rho^{n+1}_{i+\frac{1}{2},j+\frac{1}{2}}\left[u^{n+1}_{i,j+\frac{1}{2}}\right]^{-}.$$

$$(F^{n+1})^{Up,y}_{i+\frac{1}{2},j} = \rho^{n+1}_{i+\frac{1}{2},j-\frac{1}{2}}\left[v^{n+1}_{i+\frac{1}{2},j}\right]^{+} - \rho^{n+1}_{i+\frac{1}{2},j+\frac{1}{2}}\left[v^{n+1}_{i+\frac{1}{2},j}\right]^{-}. \quad (4.46)$$

Here $[\cdot]^{+} = \max\{\cdot,0\}$ and $[\cdot]^{-} = -\min\{\cdot,0\}$ represents respectively the positive and negative parts of the given function. The fast implicit part is solved via solving an elliptic equation of ρ. From the last two equations in (4.45), $u^{n+1}_{i,j+\frac{1}{2}}$ and $v^{n+1}_{i+\frac{1}{2},j}$ can be written as a function of ρ^{n+1} such that

$$
u_{i,j+\frac{1}{2}}^{n+1} = \frac{1}{\rho_{i,j+\frac{1}{2}}^{n+1}} \left[\rho_{i,j+\frac{1}{2}}^{*} u_{i,j+\frac{1}{2}}^{*} - \frac{a^n \Delta t}{\varepsilon^2} \frac{\rho_{i+\frac{1}{2},j+\frac{1}{2}}^{n+1} - \rho_{i-\frac{1}{2},j+\frac{1}{2}}^{n+1}}{\Delta x} \right.
$$

$$
\left. + \frac{a^n \Delta t}{\varepsilon^2} \rho_{i,j+\frac{1}{2}}^{n+1} \frac{\ln \rho_{0,i+\frac{1}{2},j+\frac{1}{2}} - \ln \rho_{0,i-\frac{1}{2},j+\frac{1}{2}}}{\Delta x} \right], \quad (4.47)
$$

and

$$
v_{i+\frac{1}{2},j}^{n+1} = \frac{1}{\rho_{i+\frac{1}{2},j}^{n+1}} \left[\rho_{i+\frac{1}{2},j}^{*} v_{i+\frac{1}{2},j}^{*} - \frac{a^n \Delta t}{\varepsilon^2} \frac{\rho_{i+\frac{1}{2},j+\frac{1}{2}}^{n+1} - \rho_{i+\frac{1}{2},j-\frac{1}{2}}^{n+1}}{\Delta y} \right.
$$

$$
\left. + \frac{a^n \Delta t}{\varepsilon^2} \rho_{i+\frac{1}{2},j}^{n+1} \frac{\ln \rho_{0,i+\frac{1}{2},j+\frac{1}{2}} - \ln \rho_{0,i+\frac{1}{2},j-\frac{1}{2}}}{\Delta y} \right]. \quad (4.48)
$$

Substitute the velocities by their values in the density equation from (4.45) (for simplicity, see the 1D discretization),

$$
\frac{\rho_{i+\frac{1}{2},j+\frac{1}{2}}^{n+1} - \rho_{i+\frac{1}{2},j+\frac{1}{2}}^{*}}{\Delta t} + (1-\alpha) \left[\frac{(F^{n+1})_{i+1,j+\frac{1}{2}}^{Up,x} - (F^{n+1})_{i,j+\frac{1}{2}}^{Up,x}}{\Delta x} \right.
$$

$$
\left. + \frac{(F^{n+1})_{i+\frac{1}{2},j+1}^{Up,y} - (F^{n+1})_{i+\frac{1}{2},j}^{Up,y}}{\Delta y} \right] = 0.
$$

$$
\frac{\rho_{i+\frac{1}{2},j+\frac{1}{2}}^{n+1} - \rho_{i+\frac{1}{2},j+\frac{1}{2}}^{*}}{\Delta t} + \frac{(1-\alpha)}{\Delta x} \left[\frac{\rho_{i+\frac{1}{2},j+\frac{1}{2}}^{n+1} - \rho_{i+\frac{3}{2},j+\frac{1}{2}}^{n+1}}{2\rho_{i+1,j+\frac{1}{2}}^{n+1}} \right| \rho_{i+1,j+\frac{1}{2}}^{*} u_{i+1,j+\frac{1}{2}}^{*}
$$

$$
- \frac{a^n \Delta t}{\varepsilon^2} \frac{\rho_{i+\frac{3}{2},j+\frac{1}{2}}^{n+1} - \rho_{i+\frac{1}{2},j+\frac{1}{2}}^{n+1}}{\Delta x} + \frac{a^n \Delta t}{\varepsilon^2} \rho_{i+1,j+\frac{1}{2}}^{n+1} \frac{\ln \rho_{0,i+\frac{3}{2},j+\frac{1}{2}} - \ln \rho_{0,i+\frac{1}{2},j+\frac{1}{2}}}{\Delta x} \right|
$$

$$
+ \left(\rho_{i+1,j+\frac{1}{2}}^{*} u_{i+1,j+\frac{1}{2}}^{*} - \frac{a^n \Delta t}{\varepsilon^2} \frac{\rho_{i+\frac{3}{2},j+\frac{1}{2}}^{n+1} - \rho_{i+\frac{1}{2},j+\frac{1}{2}}^{n+1}}{\Delta x} + \frac{a^n \Delta t}{\varepsilon^2} \rho_{i+1,j+\frac{1}{2}}^{n+1} \right.
$$

$$
\left. \frac{\ln \rho_{0,i+\frac{3}{2},j+\frac{1}{2}} - \ln \rho_{0,i+\frac{1}{2},j+\frac{1}{2}}}{\Delta x} \right)
$$

$$+ \frac{\rho_{i+\frac{1}{2},j+\frac{1}{2}}^{n+1} - \rho_{i-\frac{1}{2},j+\frac{1}{2}}^{n+1}}{2\rho_{i,j+\frac{1}{2}}^{n+1}} \left| \rho_{i,j+\frac{1}{2}}^{*} u_{i,j+\frac{1}{2}}^{*} - \frac{a^n \Delta t}{\varepsilon^2} \frac{\rho_{i+\frac{1}{2},j+\frac{1}{2}}^{n+1} - \rho_{i-\frac{1}{2},j+\frac{1}{2}}^{n+1}}{\Delta x} + \frac{a^n \Delta t}{\varepsilon^2} \rho_{i,j+\frac{1}{2}}^{n+1} \right.$$

$$\left. \frac{\ln \rho_{0,i+\frac{1}{2},j+\frac{1}{2}} - \ln \rho_{0,i-\frac{1}{2},j+\frac{1}{2}}}{\Delta x} \right|$$

$$- \left(\rho_{i,j+\frac{1}{2}}^{*} u_{i,j+\frac{1}{2}}^{*} - \frac{a^n \Delta t}{\varepsilon^2} \frac{\rho_{i+\frac{1}{2},j+\frac{1}{2}}^{n+1} - \rho_{i-\frac{1}{2},j+\frac{1}{2}}^{n+1}}{\Delta x} + \frac{a^n \Delta t}{\varepsilon^2} \rho_{i,j+\frac{1}{2}}^{n+1} \right.$$

$$\left. \left. \frac{\ln \rho_{0,i+\frac{1}{2},j+\frac{1}{2}} - \ln \rho_{0,i-\frac{1}{2},j+\frac{1}{2}}}{\Delta x} \right) \right]$$

$$+ \frac{(1-\alpha)}{\Delta y} \left[\frac{\rho_{i+\frac{1}{2},j+\frac{1}{2}}^{n+1} - \rho_{i+\frac{1}{2},j+\frac{3}{2}}^{n+1}}{2\rho_{i+\frac{1}{2},j+1}^{n+1}} \left| \rho_{i+\frac{1}{2},j+1}^{*} v_{i+\frac{1}{2},j+1}^{*} - \frac{a^n \Delta t}{\varepsilon^2} \frac{\rho_{i+\frac{1}{2},j+\frac{3}{2}}^{n+1} - \rho_{i+\frac{1}{2},j+\frac{1}{2}}^{n+1}}{\Delta y} \right. \right.$$

$$\left. + \frac{a^n \Delta t}{\varepsilon^2} \rho_{i+\frac{1}{2},j+1}^{n+1} \frac{\ln \rho_{0,i+\frac{1}{2},j+\frac{3}{2}} - \ln \rho_{0,i+\frac{1}{2},j+\frac{1}{2}}}{\Delta y} \right|$$

$$+ \left(\rho_{i+\frac{1}{2},j+1}^{*} v_{i+\frac{1}{2},j+1}^{*} - \frac{a^n \Delta t}{\varepsilon^2} \frac{\rho_{i+\frac{1}{2},j+\frac{3}{2}}^{n+1} - \rho_{i+\frac{1}{2},j+\frac{1}{2}}^{n+1}}{\Delta y} + \frac{a^n \Delta t}{\varepsilon^2} \rho_{i+\frac{1}{2},j+1}^{n+1} \right.$$

$$\left. \left. \frac{\ln \rho_{0,i+\frac{1}{2},j+\frac{3}{2}} - \ln \rho_{0,i+\frac{1}{2},j+\frac{1}{2}}}{\Delta y} \right) \right.$$

$$+ \frac{\rho_{i+\frac{1}{2},j+\frac{1}{2}}^{n+1} - \rho_{i+\frac{1}{2},j-\frac{1}{2}}^{n+1}}{2\rho_{i+\frac{1}{2},j}^{n+1}} \left| \rho_{i+\frac{1}{2},j}^{*} v_{i+\frac{1}{2},j}^{*} - \frac{a^n \Delta t}{\varepsilon^2} \frac{\rho_{i+\frac{1}{2},j+\frac{1}{2}}^{n+1} - \rho_{i+\frac{1}{2},j-\frac{1}{2}}^{n+1}}{\Delta y} \right.$$

$$\left. + \frac{a^n \Delta t}{\varepsilon^2} \rho_{i+\frac{1}{2},j}^{n+1} \frac{\ln \rho_{0,i+\frac{1}{2},j+\frac{1}{2}} - \ln \rho_{0,i+\frac{1}{2},j-\frac{1}{2}}}{\Delta y} \right|$$

$$- \left(\rho_{i+\frac{1}{2},j}^{*} v_{i+\frac{1}{2},j}^{*} - \frac{a^n \Delta t}{\varepsilon^2} \frac{\rho_{i+\frac{1}{2},j+\frac{1}{2}}^{n+1} - \rho_{i+\frac{1}{2},j-\frac{1}{2}}^{n+1}}{\Delta y} + \frac{a^n \Delta t}{\varepsilon^2} \rho_{i+\frac{1}{2},j}^{n+1} \right.$$

$$\left. \left. \frac{\ln \rho_{0,i+\frac{1}{2},j+\frac{1}{2}} - \ln \rho_{0,i+\frac{1}{2},j-\frac{1}{2}}}{\Delta y} \right) \right] = 0. \quad (4.49)$$

The previous system 4.49 of N^2 nonlinear equations is to be solved using the Newton-Raphson method. We are interested in solving the system $\mathbf{f(x)} = \mathbf{0}$ with $\mathbf{0} = [0.....0]^T$,

$$\mathbf{x} = [x_1 x_2 ... x_{N^2}]^T = \left[\rho_{\frac{3}{2},\frac{3}{2}}^{n+1} ... \rho_{\frac{3}{2},N+\frac{1}{2}}^{n+1} \rho_{\frac{5}{2},\frac{3}{2}}^{n+1} ... \rho_{\frac{5}{2},N+\frac{1}{2}}^{n+1} ....\rho_{N+\frac{1}{2},\frac{3}{2}}^{n+1} ... \rho_{N+\frac{1}{2},N+\frac{1}{2}}^{n+1} \right]^T,$$

$\mathbf{f(x)} = [f_1(\mathbf{x}) f_2(\mathbf{x}) ... f_{N^2}(\mathbf{x})]^T$, where T denotes the transpose operator.

Here, $f_{[(i-1)*N+j]}(\mathbf{x}) =$

$$\frac{\rho^{n+1}_{i+\frac{1}{2},j+\frac{1}{2}} - \rho^{*}_{i+\frac{1}{2},j+\frac{1}{2}}}{\Delta t} + \frac{(1-\alpha)}{\Delta x}\left[\frac{\rho^{n+1}_{i+\frac{1}{2},j+\frac{1}{2}} - \rho^{n+1}_{i+\frac{3}{2},j+\frac{1}{2}}}{2\rho^{n+1}_{i+1,j+\frac{1}{2}}}\Bigg|\rho^{*}_{i+1,j+\frac{1}{2}}u^{*}_{i+1,j+\frac{1}{2}}\right.$$

$$-\frac{a^n \Delta t}{\varepsilon^2}\frac{\rho^{n+1}_{i+\frac{3}{2},j+\frac{1}{2}} - \rho^{n+1}_{i+\frac{1}{2},j+\frac{1}{2}}}{\Delta x} + \frac{a^n \Delta t}{\varepsilon^2}\rho^{n+1}_{i+1,j+\frac{1}{2}}\frac{\ln\rho_{0,i+\frac{3}{2},j+\frac{1}{2}} - \ln\rho_{0,i+\frac{1}{2},j+\frac{1}{2}}}{\Delta x}\Bigg|$$

$$+\left(\rho^{*}_{i+1,j+\frac{1}{2}}u^{*}_{i+1,j+\frac{1}{2}} - \frac{a^n \Delta t}{\varepsilon^2}\frac{\rho^{n+1}_{i+\frac{3}{2},j+\frac{1}{2}} - \rho^{n+1}_{i+\frac{1}{2},j+\frac{1}{2}}}{\Delta x} + \frac{a^n \Delta t}{\varepsilon^2}\rho^{n+1}_{i+1,j+\frac{1}{2}}\right.$$

$$\left.\frac{\ln\rho_{0,i+\frac{3}{2},j+\frac{1}{2}} - \ln\rho_{0,i+\frac{1}{2},j+\frac{1}{2}}}{\Delta x}\right)$$

$$+\frac{\rho^{n+1}_{i+\frac{1}{2},j+\frac{1}{2}} - \rho^{n+1}_{i-\frac{1}{2},j+\frac{1}{2}}}{2\rho^{n+1}_{i,j+\frac{1}{2}}}\Bigg|\rho^{*}_{i,j+\frac{1}{2}}u^{*}_{i,j+\frac{1}{2}} - \frac{a^n \Delta t}{\varepsilon^2}\frac{\rho^{n+1}_{i+\frac{1}{2},j+\frac{1}{2}} - \rho^{n+1}_{i-\frac{1}{2},j+\frac{1}{2}}}{\Delta x} + \frac{a^n \Delta t}{\varepsilon^2}\rho^{n+1}_{i,j+\frac{1}{2}}$$

$$\frac{\ln\rho_{0,i+\frac{1}{2},j+\frac{1}{2}} - \ln\rho_{0,i-\frac{1}{2},j+\frac{1}{2}}}{\Delta x}\Bigg|$$

$$-\left(\rho^{*}_{i,j+\frac{1}{2}}u^{*}_{i,j+\frac{1}{2}} - \frac{a^n \Delta t}{\varepsilon^2}\frac{\rho^{n+1}_{i+\frac{1}{2},j+\frac{1}{2}} - \rho^{n+1}_{i-\frac{1}{2},j+\frac{1}{2}}}{\Delta x} + \frac{a^n \Delta t}{\varepsilon^2}\rho^{n+1}_{i,j+\frac{1}{2}}\right.$$

$$\left.\left.\frac{\ln\rho_{0,i+\frac{1}{2},j+\frac{1}{2}} - \ln\rho_{0,i-\frac{1}{2},j+\frac{1}{2}}}{\Delta x}\right)\right]$$

$$+\frac{(1-\alpha)}{\Delta y}\left[\frac{\rho^{n+1}_{i+\frac{1}{2},j+\frac{1}{2}} - \rho^{n+1}_{i+\frac{1}{2},j+\frac{3}{2}}}{2\rho^{n+1}_{i+\frac{1}{2},j+1}}\Bigg|\rho^{*}_{i+\frac{1}{2},j+1}v^{*}_{i+\frac{1}{2},j+1} - \frac{a^n \Delta t}{\varepsilon^2}\frac{\rho^{n+1}_{i+\frac{1}{2},j+\frac{3}{2}} - \rho^{n+1}_{i+\frac{1}{2},j+\frac{1}{2}}}{\Delta y}\right.$$

$$+\frac{a^n \Delta t}{\varepsilon^2}\rho^{n+1}_{i+\frac{1}{2},j+1}\frac{\ln\rho_{0,i+\frac{1}{2},j+\frac{3}{2}} - \ln\rho_{0,i+\frac{1}{2},j+\frac{1}{2}}}{\Delta y}\Bigg|$$

$$+\left(\rho^{*}_{i+\frac{1}{2},j+1}v^{*}_{i+\frac{1}{2},j+1} - \frac{a^n \Delta t}{\varepsilon^2}\frac{\rho^{n+1}_{i+\frac{1}{2},j+\frac{3}{2}} - \rho^{n+1}_{i+\frac{1}{2},j+\frac{1}{2}}}{\Delta y} + \frac{a^n \Delta t}{\varepsilon^2}\rho^{n+1}_{i+\frac{1}{2},j+1}\right.$$

$$\left.\frac{\ln\rho_{0,i+\frac{1}{2},j+\frac{3}{2}} - \ln\rho_{0,i+\frac{1}{2},j+\frac{1}{2}}}{\Delta y}\right)$$

$$+\frac{\rho^{n+1}_{i+\frac{1}{2},j+\frac{1}{2}} - \rho^{n+1}_{i+\frac{1}{2},j-\frac{1}{2}}}{2\rho^{n+1}_{i+\frac{1}{2},j}}\Bigg|\rho^{*}_{i+\frac{1}{2},j}v^{*}_{i+\frac{1}{2},j} - \frac{a^n \Delta t}{\varepsilon^2}\frac{\rho^{n+1}_{i+\frac{1}{2},j+\frac{1}{2}} - \rho^{n+1}_{i+\frac{1}{2},j-\frac{1}{2}}}{\Delta y}$$

$$+\frac{a^n \Delta t}{\varepsilon^2}\rho^{n+1}_{i+\frac{1}{2},j}\frac{\ln\rho_{0,i+\frac{1}{2},j+\frac{1}{2}} - \ln\rho_{0,i+\frac{1}{2},j-\frac{1}{2}}}{\Delta y}\Bigg|$$

$$-\left(\rho^{*}_{i+\frac{1}{2},j}v^{*}_{i+\frac{1}{2},j} - \frac{a^n \Delta t}{\varepsilon^2}\frac{\rho^{n+1}_{i+\frac{1}{2},j+\frac{1}{2}} - \rho^{n+1}_{i+\frac{1}{2},j-\frac{1}{2}}}{\Delta y} + \frac{a^n \Delta t}{\varepsilon^2}\rho^{n+1}_{i+\frac{1}{2},j}\right.$$

$$\left.\left.\frac{\ln\rho_{0,i+\frac{1}{2},j+\frac{1}{2}} - \ln\rho_{0,i+\frac{1}{2},j-\frac{1}{2}}}{\Delta y}\right)\right]. \quad (4.50)$$

After solving the system of nonlinear equations for ρ^{n+1}, $(\rho u)^{n+1}$ and $(\rho v)^{n+1}$ are recovered from the momentum equations. The full presentation of the 2D scheme is summarized by the slow step (4.44) together with the fast step (4.45).

4.4.3 The AP property for the 2D numerical scheme

As we did at the PDE level, we take the gravitational terms in the momentum equations in system (4.51) to the left-hand side and reformulate the system as follows:

$$
\begin{cases}
\dfrac{\rho^{n+1}_{i+\frac{1}{2},j+\frac{1}{2}}-\rho^n_{i+\frac{1}{2},j+\frac{1}{2}}}{\Delta t} + \alpha\left[\dfrac{(F^n)^{Up,x}_{i+1,j+\frac{1}{2}}-(F^n)^{Up,x}_{i,j+\frac{1}{2}}}{\Delta x} + \dfrac{(F^n)^{Up,y}_{i+\frac{1}{2},j+1}-(F^n)^{Up,y}_{i+\frac{1}{2},j}}{\Delta y}\right] \\[4mm]
+(1-\alpha)\left[\dfrac{(F^{n+1})^{Up,x}_{i+1,j+\frac{1}{2}}-(F^{n+1})^{Up,x}_{i,j+\frac{1}{2}}}{\Delta x} + \dfrac{(F^{n+1})^{Up,y}_{i+\frac{1}{2},j+1}-(F^{n+1})^{Up,y}_{i+\frac{1}{2},j}}{\Delta y}\right] = 0, \\[6mm]

\dfrac{\rho^{n+1}_{i,j+\frac{1}{2}}u^{n+1}_{i,j+\frac{1}{2}}-\rho^n_{i,j+\frac{1}{2}}u^n_{i,j+\frac{1}{2}}}{\Delta t} + \dfrac{\zeta^{u,x}_{i+\frac{1}{2},j+\frac{1}{2}}-\zeta^{u,x}_{i-\frac{1}{2},j+\frac{1}{2}}}{\Delta x} + \dfrac{\zeta^{u,y}_{i,j+1}-\zeta^{u,y}_{i,j}}{\Delta y} \\[3mm]
+\dfrac{1}{\varepsilon^2}\rho^n_{i,j+\frac{1}{2}}\left[D^x_{i,j+\frac{1}{2}}N^n + D^x_{i,j+\frac{1}{2}}\phi - a^n[D^x_{i,j+\frac{1}{2}}M^n - D^x_{i,j+\frac{1}{2}}\ln\rho_0]\right] \\[3mm]
+\dfrac{a^n}{\varepsilon^2}\rho^{n+1}_{i,j+\frac{1}{2}}[D^x_{i,j+\frac{1}{2}}M^{n+1} - D^x_{i,j+\frac{1}{2}}\ln\rho_0] = 0, \\[6mm]

\dfrac{\rho^{n+1}_{i+\frac{1}{2},j}v^{n+1}_{i+\frac{1}{2},j}-\rho^n_{i+\frac{1}{2},j}v^n_{i+\frac{1}{2},j}}{\Delta t} + \dfrac{\zeta^{v,x}_{i+1,j}-\zeta^{v,x}_{i,j}}{\Delta x} + \dfrac{\zeta^{v,y}_{i+\frac{1}{2},j+\frac{1}{2}}-\zeta^{v,y}_{i+\frac{1}{2},j-\frac{1}{2}}}{\Delta y} \\[3mm]
+\dfrac{1}{\varepsilon^2}\rho^n_{i+\frac{1}{2},j}\left[D^y_{i+\frac{1}{2},j}N^n + D^y_{i+\frac{1}{2},j}\phi - a^n[D^y_{i+\frac{1}{2},j}M^n - D^y_{i+\frac{1}{2},j}\ln\rho_0]\right] \\[3mm]
+\dfrac{a^n}{\varepsilon^2}\rho^{n+1}_{i+\frac{1}{2},j}[D^y_{i+\frac{1}{2},j}M^{n+1} - D^y_{i+\frac{1}{2},j}\ln\rho_0] = 0.
\end{cases}
\tag{4.51}
$$

with

$$
D^x_{i,j+\frac{1}{2}}\rho = \frac{\rho_{i+\frac{1}{2},j+\frac{1}{2}} - \rho_{i-\frac{1}{2},j+\frac{1}{2}}}{\Delta x}, \qquad D^y_{i+\frac{1}{2},j}\rho = \frac{\rho_{i+\frac{1}{2},j+\frac{1}{2}} - \rho_{i+\frac{1}{2},j-\frac{1}{2}}}{\Delta y}.
$$

We will show the AP property based on (4.51).

Assume that the Chapman-Enskog asymptotic expansion of the discrete variables are

$$
\rho^n_{i+\frac{1}{2},j+\frac{1}{2}} = \rho^{(0)n}_{i+\frac{1}{2},j+\frac{1}{2}} + \varepsilon\rho^{(1)n}_{i+\frac{1}{2},j+\frac{1}{2}} + \varepsilon^2\rho^{(2)n}_{i+\frac{1}{2},j+\frac{1}{2}} + \cdots,
$$

$$
u^n_{i,j+\frac{1}{2}} = u^{(0)n}_{i,j+\frac{1}{2}} + \varepsilon u^{(1)n}_{i,j+\frac{1}{2}} + \varepsilon^2 u^{(2)n}_{i,j+\frac{1}{2}} + \cdots,
$$

$$
v^n_{i+\frac{1}{2},j} = v^{(0)n}_{i+\frac{1}{2},j} + \varepsilon v^{(1)n}_{i+\frac{1}{2},j} + \varepsilon^2 v^{(2)n}_{i+\frac{1}{2},j} + \cdots.
$$

Definition 2. *The discrete data (ρ, u, v) are said to be well-prepared if they satisfy,*

$$
\rho^n_{i+\frac{1}{2},j+\frac{1}{2}} = \left(1 - \frac{\gamma-1}{\gamma A}\phi_{i+\frac{1}{2},j+\frac{1}{2}}\right)^{\frac{1}{\gamma-1}} + O(\varepsilon^2) = \rho^{(0)}_{i+\frac{1}{2},j+\frac{1}{2}} + O(\varepsilon^2)
$$

$$
\frac{(F^{(0)n})^{Up,x}_{i+1,j+\frac{1}{2}} - (F^{(0)n})^{Up,x}_{i,j+\frac{1}{2}}}{\Delta x} + \frac{(F^{(0)n})^{Up,y}_{i+\frac{1}{2},j+1} - (F^{(0)n})^{Up,y}_{i+\frac{1}{2},j}}{\Delta y} = 0,
$$

$$
\tag{4.52}
$$

where $(F^{(0),n})^{Up,x}$, $(F^{(0),n})^{Up,y}$ are defined as in (4.46) with ρ^{n+1}, u^{n+1} being replaced by $\rho^{(0),n}$, $u^{(0),n}$.

Lemma 3. *Choose (ρ, u, v) to be well-prepared, then*

$$\mathcal{L}_d = \frac{1}{\varepsilon^2}\rho^n_{i,j+\frac{1}{2}}\left[D^x_{i,j+\frac{1}{2}}N^n + D^x_{i,j+\frac{1}{2}}\phi - a^n_d[D^x_{i,j+\frac{1}{2}}M^n - D^x_{i,j+\frac{1}{2}}\ln\rho_0]\right]$$

$$= \rho^{(0)}_{i,j+\frac{1}{2}}D^x_{i,j+\frac{1}{2}}\left[N^{n(2)} - a^{(0)}M^{n(2)}\right] + O(\varepsilon).$$

Proof. It is easy to check that

$$D^x_{i,j+\frac{1}{2}}M(\rho^{(0)}) - D^x_{i,j+\frac{1}{2}}\ln\rho_0 = 0, \qquad D^y_{i+\frac{1}{2},j}M(\rho^{(0)}) - D^y_{i+\frac{1}{2},j}\ln\rho_0 = 0,$$

$$D^x_{i,j+\frac{1}{2}}N(\rho^{(0)}) = \frac{A\gamma}{\gamma-1}D^x_{i,j+\frac{1}{2}}(\rho^{(0)})^{\gamma-1} = -D^x_{i,j+\frac{1}{2}}\phi, \quad D^y_{i+\frac{1}{2},j}N(\rho^{(0)}) + D^y_{i+\frac{1}{2},j}\phi = 0,$$

$$\left(D^x_{i,j+\frac{1}{2}}M^n\right)^{(1)} = \left(D^x_{i,j+\frac{1}{2}}N^n\right)^{(1)} = 0, \qquad \left(D^y_{i+\frac{1}{2},j}M^n\right)^{(1)} = \left(D^y_{i+\frac{1}{2},j}N^n\right)^{(1)} = 0.$$

$$\tag{4.53}$$

The expansion of a^n is the same as in (4.25), (4.26). Now, let's look at the expansion of $\mathcal{L}_d$. Noting $a^{(0)n} = a^{(0)}$ and (4.53), we have

$$\mathcal{L}_d = \frac{1}{\varepsilon^2}\rho^{(0)}_{i,j+\frac{1}{2}}\left[(D^x_{i,j+\frac{1}{2}}N^n)^{(0)} + D^x_{i,j+\frac{1}{2}}\phi - a^{(0)n}[(D^x_{i,j+\frac{1}{2}}M^n)^{(0)} - D^x_{i,j+\frac{1}{2}}\ln\rho_0]\right]$$

$$+ \rho^{(0)}_{i,j+\frac{1}{2}}\left[(D^x_{i,j+\frac{1}{2}}N^n)^{(2)} - a^{(0)n}(D^x_{i,j+\frac{1}{2}}M^n)^{(2)}\right.$$

$$\left. - a^{(2)n}[(D^x_{i,j+\frac{1}{2}}M^n)^{(0)} - D^x_{i,j+\frac{1}{2}}\ln\rho_0]\right]$$

$$+ \rho^{(2)n}_{i,j+\frac{1}{2}}\left[(D^x_{i,j+\frac{1}{2}}N^n)^{(0)} + D^x_{i,j+\frac{1}{2}}\phi\right.$$

$$\left. - a^{(0)n}[(D^x_{i,j+\frac{1}{2}}M^n)^{(0)} - D^x_{i,j+\frac{1}{2}}\ln\rho_0]\right] + O(\varepsilon),$$

$$= \rho^{(0)n}_{i,j+\frac{1}{2}}D^x_{i,j+\frac{1}{2}}\left[(N^n)^{(2)} - a^{(0)}(M^n)^{(2)}\right] + O(\varepsilon),$$

which conclude the proof. □

Similarly,

$$\frac{1}{\varepsilon^2}\rho^n_{i+\frac{1}{2},j}\left[D^y_{i+\frac{1}{2},j}N^n + D^y_{i+\frac{1}{2},j}\phi - a^n_d[D^y_{i+\frac{1}{2},j}M^n - D^y_{i+\frac{1}{2},j}\ln\rho_0]\right]$$

$$= \rho^{(0)n}_{i+\frac{1}{2},j}D^x_{i+\frac{1}{2},j}\left[(N^n)^{(2)} - a^{(0)}(M^n)^{(2)}\right] + O(\varepsilon).$$

Comparing $O(\frac{1}{\varepsilon^2})$ terms in the momentum equation in the x-direction, one gets

$$a^{(0),n}_d\rho^{(0),n+1}_{i,j+\frac{1}{2}}[(D^x_{i,j+\frac{1}{2}}M^{n+1})^{(0)} - D^x_{i,j+\frac{1}{2}}\ln\rho_0] = 0. \tag{4.54}$$

Because $a^{(0),n}_d \neq 0$ and $\rho^{(0),n+1}_{i,j+\frac{1}{2}} \neq 0$, then

$$D^x_{i,j+\frac{1}{2}}\ln\rho^{(0)n+1} = -D^x_{i,j+\frac{1}{2}}\ln(1 - \frac{\gamma-1}{\gamma A}\phi)^{\frac{1}{\gamma-1}}.$$

Hence, the boundary conditions of ρ^{n+1} yield

$$\rho^{(0),n+1}_{i+\frac{1}{2},j+\frac{1}{2}} = \left(1 - \frac{\gamma - 1}{\gamma A}\phi_{i+\frac{1}{2},j+\frac{1}{2}}\right)^{\frac{1}{\gamma-1}}$$

Similar result can be obtained from comparing $\mathcal{O}(\frac{1}{\varepsilon^2})$ terms in the momentum equation in the y-direction. From the above calculations, we deduce that $\rho^{(0)}_{i+\frac{1}{2},j+\frac{1}{2}}$ is independent of time. Similarly, comparing $\mathcal{O}(\frac{1}{\varepsilon})$ terms in the momentum equation gives $\rho^{(1)}_{i+\frac{1}{2},j+\frac{1}{2}} = 0$. From $\mathcal{O}(1)$ terms in the density equation and that $\rho^{(0)}_{i+\frac{1}{2},j+\frac{1}{2}}$ is time independent,

$$\alpha \left[\frac{(F^{n(0)})^x_{i+1,j+\frac{1}{2}} - (F^{n(0)})^x_{i,j+\frac{1}{2}}}{\Delta x} + \frac{(F^{n(0)})^y_{i+\frac{1}{2},j+1} - (F^{n(0)})^y_{i+\frac{1}{2},j}}{\Delta y} \right]$$

$$+ (1-\alpha) \left[\frac{(F^{n+1(0)})^{Up,x}_{i+1,j+\frac{1}{2}} - (F^{n+1(0)})^{Up,x}_{i,j+\frac{1}{2}}}{\Delta x} \right.$$

$$\left. + \frac{(F^{n+1(0)})^{Up,y}_{i+\frac{1}{2},j+1} - (F^{n+1(0)})^{Up,y}_{i+\frac{1}{2},j}}{\Delta y} \right] = 0.$$

Substitute the fluxes by their values,

$$\alpha \left[\frac{(\rho^{(0)}u^{(0)})^n_{i+1,j+\frac{1}{2}} - (\rho^{(0)}u^{(0)})^n_{i,j+\frac{1}{2}}}{\Delta x} + \frac{(\rho^{(0)}v^{(0)})^n_{i+\frac{1}{2},j+1} - (\rho^{(0)}u^{(0)})^n_{i+\frac{1}{2},j}}{\Delta y} \right]$$

$$+ (1-\alpha) \left[\frac{(\rho^{(0)}u^{(0)})^{n+1}_{i+1,j+\frac{1}{2}} - (\rho^{(0)}u^{(0)})^{n+1}_{i,j+\frac{1}{2}}}{\Delta x} \right.$$

$$\left. + \frac{(\rho^{(0)}v^{(0)})^{n+1}_{i+\frac{1}{2},j+1} - (\rho^{(0)}v^{(0)})^{n+1}_{i+\frac{1}{2},j}}{\Delta y} \right] = 0.$$

The discrete well-prepared initial data in (4.52) lead to,

$$\frac{(\rho^{(0)}u^{(0)})^{n+1}_{i+1,j+\frac{1}{2}} - (\rho^{(0)}u^{(0)})^{n+1}_{i,j+\frac{1}{2}}}{\Delta x}$$

$$+ \frac{(\rho^{(0)}v^{(0)})^{n+1}_{i+\frac{1}{2},j+1} - (\rho^{(0)}u^{(0)})^{n+1}_{i+\frac{1}{2},j}}{\Delta y} = 0, \quad \forall(i,j). \quad (4.55)$$

Comparing order $\mathcal{O}(1)$ terms in the momentum equation in x-direction,

$$\frac{\rho^{(0)}_{i,j+\frac{1}{2}} u^{n+1(0)}_{i,j+\frac{1}{2}} - \rho^{(0)}_{i,j+\frac{1}{2}} u^{n(0)}_{i,j+\frac{1}{2}}}{\Delta t} + \frac{\rho^{(0)}_{i,j+\frac{1}{2}} \zeta^{u,x}_{i+\frac{1}{2},j+\frac{1}{2}} - \rho^{(0)}_{i,j+\frac{1}{2}} \zeta^{u,x}_{i-\frac{1}{2},j+\frac{1}{2}}}{\Delta x}$$

$$+ \frac{\rho^{(0)}_{i,j+\frac{1}{2}} \zeta^{u,y}_{i,j+1} - \rho^{(0)}_{i,j+\frac{1}{2}} \zeta^{u,y}_{i,j}}{\Delta y} + \rho^{(0)n}_{i,j+\frac{1}{2}} D^x_{i,j+\frac{1}{2}} \left[(N^n)^{(2)} - a^{(0)}(M^n)^{(2)} \right]$$

$$+ a^{(0)n}_d \rho^{(0)n+1}_{i,j+\frac{1}{2}} (D^x_{i,j+\frac{1}{2}} M^{n+1})^{(2)} + (a^{(0)n}_d \rho^{(1)n+1}_{i,j+\frac{1}{2}} + a^{(1)n}_d \rho^{(0)n+1}_{i,j+\frac{1}{2}})(D^x_{i,j+\frac{1}{2}} M^{n+1})^{(1)}$$

$$+ (a^{(0)n}_d \rho^{(2)n+1}_{i,j+\frac{1}{2}} + a^{(1),n}_d \rho^{(1)}_{i,j+\frac{1}{2}} + a^{(2)n}_d \rho^{(0)n+1}_{i,j+\frac{1}{2}})\left[(D^x_{i,j+\frac{1}{2}} M^{n+1})^{(0)} - D^x_{i,j+\frac{1}{2}} \ln \rho_0 \right] = 0.$$

Using (4.54) the equation can be simplified to,

$$\frac{u^{n+1(0)}_{i,j+\frac{1}{2}} - u^{n(0)}_{i,j+\frac{1}{2}}}{\Delta t} + \frac{\zeta^{u,x}_{i+\frac{1}{2},j+\frac{1}{2}} - \zeta^{u,x}_{i-\frac{1}{2},j+\frac{1}{2}}}{\Delta x} + \frac{\zeta^{u,y}_{i,j+1} - \zeta^{u,y}_{i,j}}{\Delta y}$$

$$+ D^x_{i,j+\frac{1}{2}} \left[(N^n)^{(2)} - a^{(0)}(M^n)^{(2)} + a^{(0)}(M^{n+1})^{(2)} \right] = 0.$$

Finally the momentum limit equation in the x-direction is,

$$\frac{u^{n+1(0)}_{i,j+\frac{1}{2}} - u^{n(0)}_{i,j+\frac{1}{2}}}{\Delta t} + \frac{\zeta^{u,x}_{i+\frac{1}{2},j+\frac{1}{2}} - \zeta^{u,x}_{i-\frac{1}{2},j+\frac{1}{2}}}{\Delta x}$$

$$+ \frac{\zeta^{u,y}_{i,j+1} - \zeta^{u,y}_{i,j}}{\Delta y} + \frac{W^{(2)n+1}_{i+\frac{1}{2},j+\frac{1}{2}} - W^{(2)n+1}_{i-\frac{1}{2},j+\frac{1}{2}}}{\Delta x} = 0. \quad (4.56)$$

Similar calculations are performed on the momentum equation in the y-direction, which yields

$$\frac{v^{n+1(0)}_{i+\frac{1}{2},j} - v^{n(0)}_{i+\frac{1}{2},j}}{\Delta t} + \frac{\zeta^{u,x}_{i+\frac{1}{2},j+\frac{1}{2}} - \zeta^{u,x}_{i-\frac{1}{2},j+\frac{1}{2}}}{\Delta x}$$

$$+ \frac{\zeta^{u,y}_{i,j+1} - \zeta^{u,y}_{i,j}}{\Delta y} + \frac{W^{(2)n+1}_{i+\frac{1}{2},j+\frac{1}{2}} - W^{(2)n+1}_{i+\frac{1}{2},j-\frac{1}{2}}}{\Delta y} = 0. \quad (4.57)$$

Therefore, the fully-discrete incompressible limit system is:

$$\begin{cases} \frac{(\rho^{(0)}u^{(0)})^{n+1}_{i+1,j+\frac{1}{2}} - (\rho^{(0)}u^{(0)})^{n+1}_{i,j+\frac{1}{2}}}{\Delta x} + \frac{(\rho^{(0)}v^{(0)})^{n+1}_{i+\frac{1}{2},j+1} - (\rho^{(0)}u^{(0)})^{n+1}_{i+\frac{1}{2},j}}{\Delta y} = 0, \\[2mm] \frac{u^{n+1(0)}_{i,j+\frac{1}{2}} - u^{n(0)}_{i,j+\frac{1}{2}}}{\Delta t} + \frac{\zeta^{u,x}_{i+\frac{1}{2},j+\frac{1}{2}} - \zeta^{u,x}_{i-\frac{1}{2},j+\frac{1}{2}}}{\Delta x} + \frac{\zeta^{u,y}_{i,j+1} - \zeta^{u,y}_{i,j}}{\Delta y} + \frac{W^{(2)n+1}_{i+\frac{1}{2},j+\frac{1}{2}} - W^{(2)n+1}_{i-\frac{1}{2},j+\frac{1}{2}}}{\Delta x} = 0, \\[2mm] \frac{v^{n+1(0)}_{i+\frac{1}{2},j} - v^{n(0)}_{i+\frac{1}{2},j}}{\Delta t} + \frac{\zeta^{u,x}_{i+\frac{1}{2},j+\frac{1}{2}} - \zeta^{u,x}_{i-\frac{1}{2},j+\frac{1}{2}}}{\Delta x} + \frac{\zeta^{u,y}_{i,j+1} - \zeta^{u,y}_{i,j}}{\Delta y} + \frac{W^{(2)n+1}_{i+\frac{1}{2},j+\frac{1}{2}} - W^{(2)n+1}_{i+\frac{1}{2},j-\frac{1}{2}}}{\Delta y} = 0. \end{cases}$$

$$(4.58)$$

4.4.4   The SP property of the 2D numerical scheme

In this section we prove that the developed AP scheme for the isentropic Euler equations with gravitational source term is SP. The stationary state under consideration is

$$u^s_{i,j+\frac{1}{2}} = v^s_{i+\frac{1}{2},j} = 0, \qquad \rho^s_{i+\frac{1}{2},j+\frac{1}{2}} = \left(1 - \frac{\gamma - 1}{\gamma A}\phi_{i+\frac{1}{2},j+\frac{1}{2}}\right)^{\frac{1}{\gamma-1}}.$$

Theorem 2. *If the solution at time t^n is stationary,*
i.e. $(\rho^n_{i+\frac{1}{2},j+\frac{1}{2}}, u^n_{i,j+\frac{1}{2}}, v^n_{i+\frac{1}{2},j}) = (\rho^s_{i+\frac{1}{2},j+\frac{1}{2}}, u^s_{i,j+\frac{1}{2}}, v^s_{i+\frac{1}{2},j})$, *then it does not change at the next time t^{n+1}.*

Proof.   By substituting $(\rho^n_{i+\frac{1}{2},j+\frac{1}{2}}, u^n_{i,j+\frac{1}{2}}, v^n_{i+\frac{1}{2},j}) = (\rho^s_{i+\frac{1}{2},j+\frac{1}{2}}, u^s_{i,j+\frac{1}{2}}, v^s_{i+\frac{1}{2},j})$ into (4.51), it is easy to check that

$$D^x_{i,j+\frac{1}{2}} N^n + D^x_{i,j+\frac{1}{2}}\phi = 0, \qquad D^y_{i+\frac{1}{2},j} N^n + D^y_{i+\frac{1}{2},j}\phi = 0,$$
$$D^x_{i,j+\frac{1}{2}} M^n - D^x_{i,j+\frac{1}{2}}\ln\rho_0 = 0, \qquad D^y_{i+\frac{1}{2},j} M^n - D^y_{i+\frac{1}{2},j}\ln\rho_0 = 0,$$

and for $\forall\varepsilon$, (4.51) becomes

$$\begin{cases} \dfrac{\rho^{n+1}_{i+\frac{1}{2},j+\frac{1}{2}} - \rho^n_{i+\frac{1}{2},j+\frac{1}{2}}}{\Delta t} + (1-\alpha)\left[\dfrac{(F^{n+1})^{Up,x}_{i+1,j+\frac{1}{2}} - (F^{n+1})^{Up,x}_{i,j+\frac{1}{2}}}{\Delta x}\right. \\ \left. + \dfrac{(F^{n+1})^{Up,y}_{i+\frac{1}{2},j+1} - (F^{n+1})^{Up,y}_{i+\frac{1}{2},j}}{\Delta y}\right] = 0, \\[2mm] \dfrac{\rho^{n+1}_{i,j+\frac{1}{2}} u^{n+1}_{i,j+\frac{1}{2}}}{\Delta t} + \dfrac{a^n}{\varepsilon^2}\rho^{n+1}_{i,j+\frac{1}{2}}[D^x_{i,j+\frac{1}{2}} M^{n+1} - D^x_{i,j+\frac{1}{2}}\ln\rho_0] = 0, \\[2mm] \dfrac{\rho^{n+1}_{i+\frac{1}{2},j} v^{n+1}_{i+\frac{1}{2},j}}{\Delta t} + \dfrac{a^n}{\varepsilon^2}\rho^{n+1}_{i+\frac{1}{2},j}[D^y_{i+\frac{1}{2},j} M^{n+1} - D^y_{i+\frac{1}{2},j}\ln\rho_0] = 0. \end{cases} \tag{4.59}$$

One can check that $(\rho^{n+1}_{i+\frac{1}{2},j+\frac{1}{2}}, u^{n+1}_{i,j+\frac{1}{2}}, v^{n+1}_{i+\frac{1}{2},j}) = (\rho^s_{i+\frac{1}{2},j+\frac{1}{2}}, u^s_{i,j+\frac{1}{2}}, v^s_{i+\frac{1}{2},j})$ satisfies (4.59), this concludes the proof of the SP property of the fully-discretized scheme.   □

4.5   Numerical Results

We validate the 1D and 2D numerical schemes in this section, with an interest in the AP and the SP property of both schemes. Experiments are chosen for the isentropic Euler equations with and without gravitational source term. Note that in the absence of the gravitational source term, the scheme reduces to the AP scheme developed by Goudon et al. [34]. As in [34], we choose $\alpha = \varepsilon^2$ and $l = 0$ in the definition of $a(t)$ for all numerical experiments.

4.5.1   1D test cases

1D Riemann problem

To validate the robustness of the numerical scheme, we extract from [34] a 1D Riemann problem for different values of ε. The initial conditions are

$$\rho(x,0) = \begin{cases} 1 + \varepsilon^2 & \text{if } x < 0.5, \\ 1 & \text{if } x > 0.5, \end{cases}$$

$$u(x,0) = \begin{cases} 1 - \varepsilon & \text{if } x < 0.5, \\ 1 + \varepsilon & \text{if } x > 0.5. \end{cases}$$

The pressure is given by $p(\rho) = A\rho^\gamma$ with $A = 1$ and $\gamma = 2$. The solution is computed along the interval $[0,1]$ over 200 grid points for $\delta t = \beta \delta x$, with $\beta = 0.2, 0.1$ or 0.01. To test the AP property of the scheme, three differenet cases for different values of ε are considered. The density and the velocity are illustrated at the final time $T = 0.1$ for $\varepsilon = \sqrt{0.99}$ and $\beta = 0.2$ in figure 4.2, at the final time $T = 0.05$ for $\varepsilon = \sqrt{0.1}$ and $\beta = 0.1$ in figure 4.3, and at the final time $T = 0.007$ for $\varepsilon = \sqrt{0.001}$ and $\beta = 0.01$ in figure 4.4. Note that in the cases where ε is small ($\varepsilon = \sqrt{0.1}$ or $\sqrt{0.001}$), the AP scheme gives relevant results for $\beta = 0.2$, while explicit scheme simply returns negative density. By adjusting β, the AP scheme gives better results, and the explicit scheme returns positive density. For more details about the comparison, please refer to section 3.1 in [34]. The plots are in prefect match with the ones in the Literature. The solution can still be captured as ε gets smaller which proves the AP property of the 1D scheme (4.35)-(4.36).

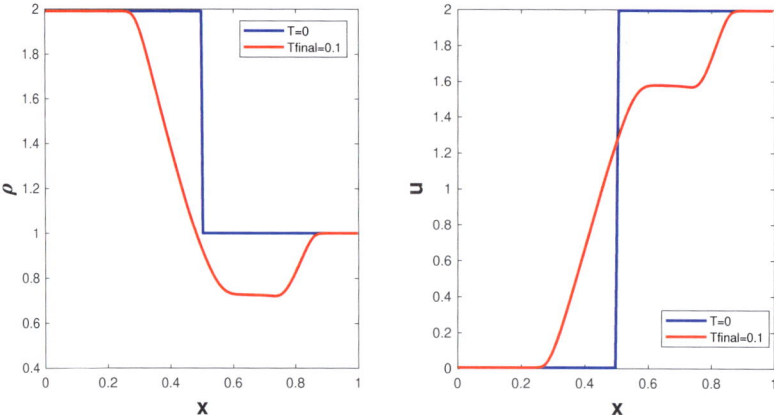

Figure 4.2: 1D Riemann problem: density (left) and velocity (right) initially, and at the final time Tfinal=0.1 for $\varepsilon = \sqrt{0.99}$ and $\beta = 0.2$.

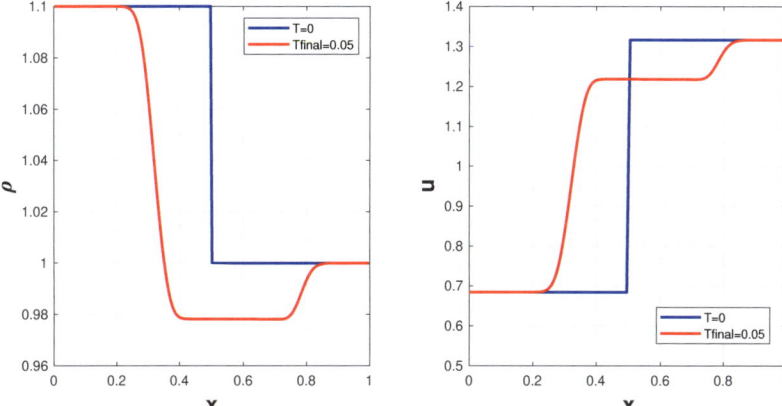

Figure 4.3: 1D Riemann problem: density (left) and velocity (right) initially, and at the final time Tfinal=0.05 for $\varepsilon = \sqrt{0.1}$ and $\beta = 0.1$.

1D steady state

As proven analytically, the AP scheme is also SP. For this purpose, we try to simulate a steady state solution, and prove numerically that the scheme preserves such a state. One example of a steady state for the isentropic Euler equations with gravitational source term is

$$\begin{cases} \rho(x) = \left(A(1 - \frac{\gamma-1}{\gamma} \frac{1}{A}\phi(x)) \right)^{\frac{1}{\gamma-1}}, \\ u(x) = 0. \end{cases} \tag{4.60}$$

With the pressure law given as $p(\rho) = A\rho^{\gamma}$ where $A = 1$ and $\gamma = 1.4$, and a gravitational potential $\phi(x) = x$. At the PDE level, (4.60) is a steady state solution. The computational domain is the interval $[0, 1]$ discretized over 200 grid points. We choose $\varepsilon = \sqrt{0.99}$ and $\delta t = \beta \delta x$ with $\beta = 0.01$. With the knowledge that the scheme should preserve the steady state independent of the choice of ε. We run our simulations till the final time $t = 0.1$ and compare it to the steady state solution in figure 4.5. The density plot at the final time lies exactly on top of the initial density. The velocity error is approximately 10^{-7} and this error stays as it is as time increases, an indication that the scheme has reached the numerical steady state. It is worth mentioning that no well-balancing treatment is applied here. In other words, the AP schemes with their IMEX structure fulfill the need for any SP treatment. At least for the isentropic Euler equations with gravitational source term, the SP property follows from the AP property.

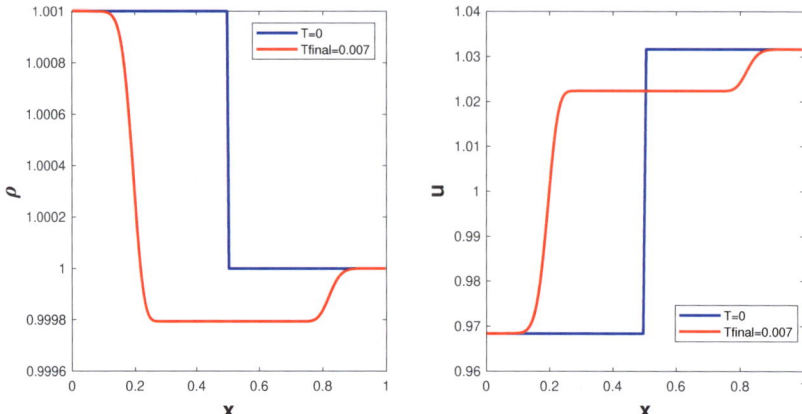

Figure 4.4: 1D Riemann problem: density (left) and velocity (right) initially, and at the final time Tfinal=0.007 for $\varepsilon = \sqrt{0.001}$ and $\beta = 0.01$.

4.5.2 2D test cases

2D Riemann problem

An extension of the 1D Riemann Problem is considered in this section. The initial data are given as

$$\rho(x, y, 0) = \begin{cases} 1 + \varepsilon^2 & \text{if } x < 0.5, \\ 1 & \text{if } x > 0.5, \end{cases}$$

$$u(x, y, 0) = \begin{cases} 1 - \varepsilon & \text{if } x < 0.5, \\ 1 + \varepsilon & \text{if } x > 0.5, \end{cases}$$

$$v(x, y, 0) = 0.$$

The 1D flow in 2D setup takes place in the direction of the horizontal velocity. The computational domain is the square $(0, 1) \times (0, 1)$ divided into 200×200 grid points. A comparison between the 1D results and the 2D cross sections is illustrated. The density and the velocity are plotted at the final time $T = 0.1$ for $\varepsilon = \sqrt{0.99}$ and $\beta = 0.2$ in figure 4.6, at the final time $T = 0.05$ for $\varepsilon = \sqrt{0.1}$ and $\beta = 0.1$ in figure 4.7, and at the final time $T = 0.007$ for $\varepsilon = \sqrt{0.001}$ and $\beta = 0.01$ in figure 4.8. The results show the accuracy and the robustness of the 2D scheme (4.44)-(4.45) as well as the AP property.

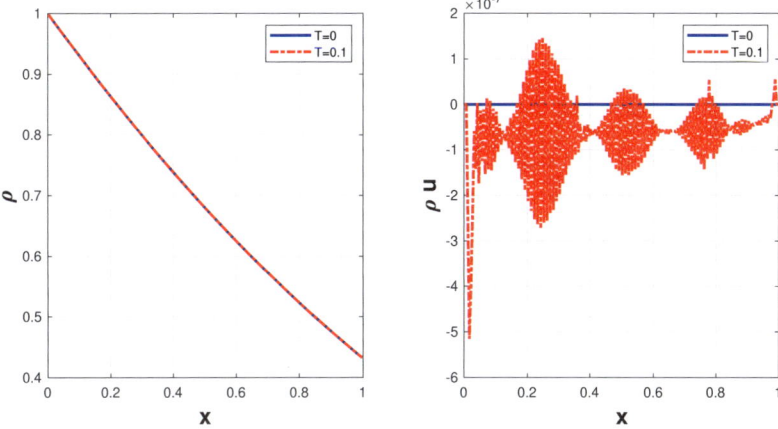

Figure 4.5: 1D steady state: profile of the density (left) and the momentum (right) initially and at the final time $t = 0.1$

2D steady state

In this section, we test the SP property of the 2D scheme. An extension of the 1D steady state along the y-axis is considered

$$\rho(x, y) = \left(A(1 - \frac{\gamma - 1}{\gamma}\frac{1}{A}\phi(x, y)) \right)^{\frac{1}{\gamma - 1}}, \qquad (4.61)$$

$$(4.62)$$

with zero velocity field $\mathbf{u} = 0$ in the square $(0, 1) \times (0, 1)$, over 200×200 grid points, and a gravitational potential $\phi(x, y) = x$. A direct comparison between the 1D plots and the 2D cross sections is illustrated in figure 4.9. This test case proves that the 2D AP scheme preserves steady states numerically without the need for any extra well-balancing, which is a strong statement, suggesting that we can prove, so far (analytically and numerically), for AP schemes for the isentropic Euler equations with gravitational source term.

2D translating vortex

A traveling vortex from [34] is considered in this section. The computational domain is the square $[0, 1] \times [0, 1]$ discretized over 32×32 grid points with $\varepsilon = 0.8$ and

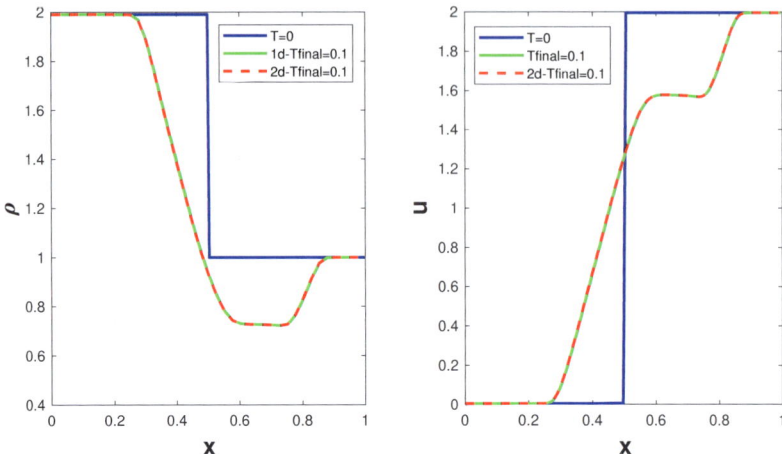

Figure 4.6: 2D Riemann problem: density (left) and velocity (right) initially, and at the final time Tfinal=0.1 for $\varepsilon = \sqrt{0.99}$ and $\beta = 0.2$.

$\delta t = 5 \times 10^{-4}$. The initial data are given as

$$\rho(x, y, 0) = 110 + \frac{\varepsilon^2}{(4\pi)^2} f(r),$$
$$u(x, y, 0) = \nu_0 + g(r)(0.5 - y),$$
$$v(x, y, 0) = \nu_1 + g(r)(x - 0.5),$$

with

$$r = 4\pi((x - 0.5)^2 + (y - 0.5)^2))^{\frac{1}{2}},$$
$$f(r) = (1.5)^2 \delta(r)(k(r) - k(\pi)),$$
$$g(r) = 1.5(1 + \cos(r))\delta(r),$$
$$\delta(r) = 1_{r<\pi}.$$

The pressure law is given as $p(\rho) = \frac{1}{2}\rho^2$ and $\nu_0 = 0.6, \nu_1 = 0$. We compare our computed numerical solution to the exact solution,

$$\rho(x, y, t) = \rho(x - \nu_0 t, y - \nu_1 t, 0),$$
$$u(x, y, 0) = u(x - \nu_0 t, y - \nu_1 t, 0),$$
$$v(x, y, 0) = v(x - \nu_0 t, y - \nu_1 t, 0).$$

The vortex gets translated at speed (ν_0, ν_1), as one can see in figures 4.10 and 4.11. We present initially and at the final time, the horizontal velocity in figures 4.12 and

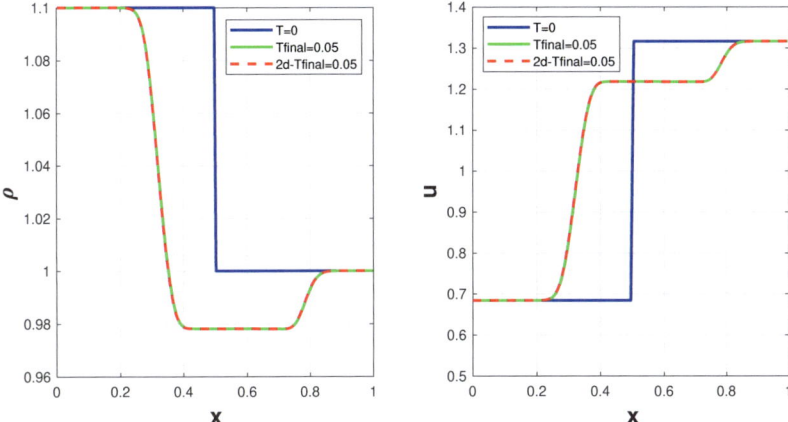

Figure 4.7: 2D Riemann problem: density (left) and velocity (right) initially, and at the final time Tfinal=0.05 for $\varepsilon = \sqrt{0.1}$ and $\beta = 0.1$.

4.13, and the vertical velocity in figures 4.14 and 4.15. To avoid spurious oscillation, we set l in definition of $a(t)$ to 1.

2D stationary vortex

For our last test case, we consider a stationary vortex for the system of isentropic Euler equations with gravitational source term. The aim is to prove that our numerical scheme is both SP, as for a fixed ε, the vortex is a stationary solution of the system and AP, as the numerical solution becomes a solution of the incompressible version of the isentropic Euler system as ε goes to zero. We take the vortex for the shallow water equations defined in [59], and we change its initial data to fit the the rescaled shallow water equations. The initial condistions are given as,

$$\rho(x, y, t) = 1 - \frac{\varepsilon^2}{4} e^{2(1-r^2)} - \phi(x, y), \quad u(t, x, y) = ye^{1-r^2}, \quad v(t, x, y) = -xe^{1-r^2}.$$

Here $r^2 = x^2 + y^2$, $\phi(x, y) = 0.2e^{0.5(1-r^2)}$ is the gravitational potential. The pressure law is $p(\rho) = A\rho^\gamma$ with $A = \frac{1}{2}$ and $\gamma = 2$. The vortex rotates in the computational domain $(-1, 1)$ x $(-1, 1)$ with steady state boundary conditions over 32 x 32 grid points. Figure 4.16 illustrates the profile of the velocity $q = \sqrt{u^2 + v^2}$ initially and figures 4.17-4.18 at the final time $t = 1$ for $\varepsilon = 10^{-1}, 10^{-2}, 10^{-3}, 10^{-4}$ respectively. The significance of this test case lies in the fact that the scheme preserves the steady state and at the same time converges as ε goes to zero. The result ensures the ability of our numerical scheme to preserve steady states and to capture the solution as ε gets smaller. This test case proves that the developed numerical scheme for the system of isentropic Euler equations with gravitational source term is both SP and AP.

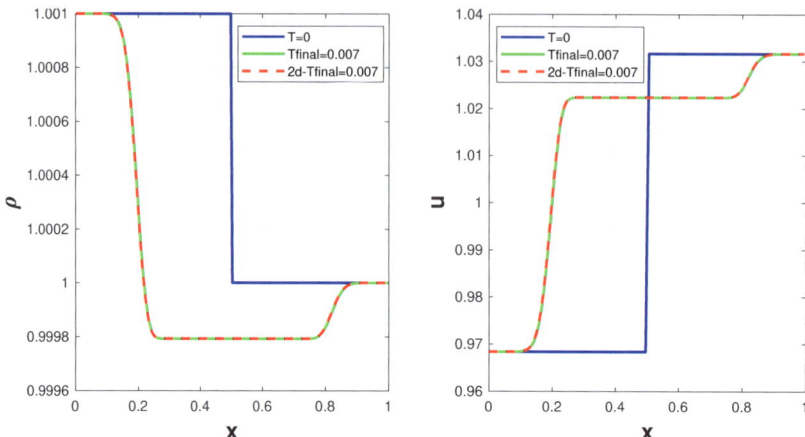

Figure 4.8: 2D Riemann problem: density (left) and velocity (right) initially, and at the final time Tfinal=0.007 for $\varepsilon = \sqrt{0.001}$ and $\beta = 0.01$.

4.6 Conclusion

The proof of the SP property at the semi-discrete level clearly depends on the pressure law and the fact that we are in the isentropic case. An AP scheme for the isentropic Euler equations is SP under the condition that the pressure is a function of the density and that the latter is obtained as a solution of an elliptic equation [50].

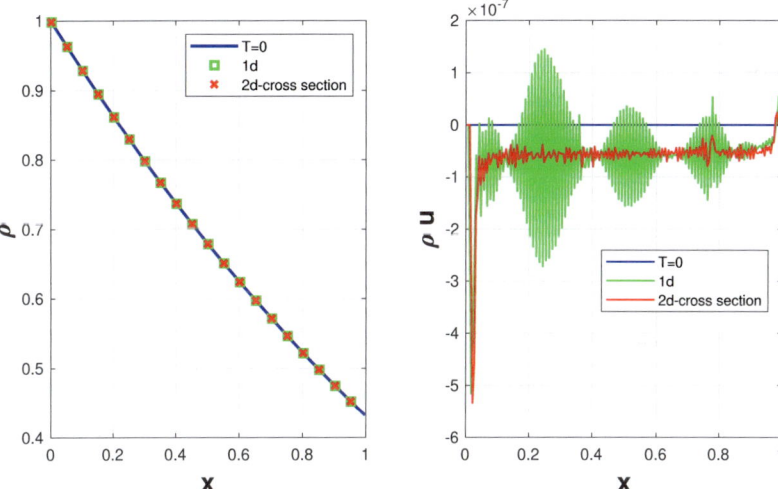

Figure 4.9: 2D steady state: profile of the density (left) and the momentum (right) initially and at the final time $t = 0.1$.

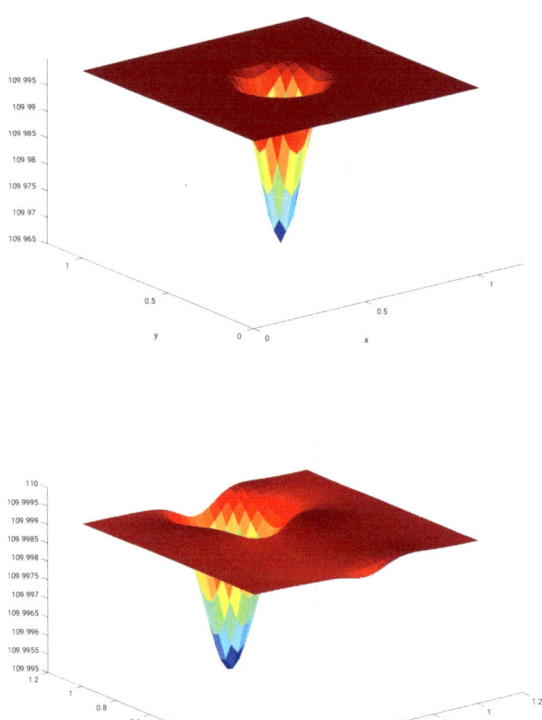

Figure 4.10: Translating vortex: the initial density ρ at T $= 0$ (top) and at the final time T $= 0.5$ (bottom).

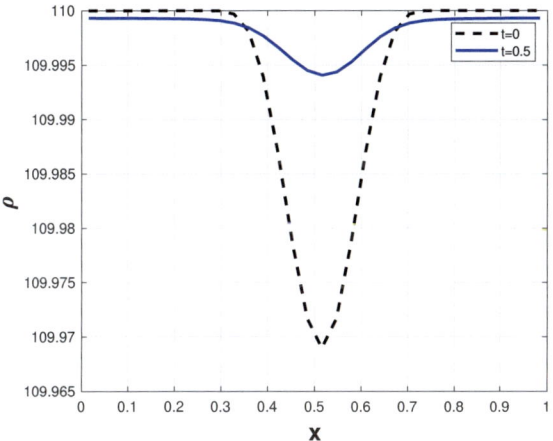

Figure 4.11: Translating vortex: a cross section of the the initial density ρ at T $= 0$ with $\min_{i,j} \rho_{i+\frac{1}{2},j+\frac{1}{2}} = 109.9690$ and at the final time T $= 0.5$ with $\min_{i,j} \rho_{i+\frac{1}{2},j+\frac{1}{2}} = 109.9951$ along $y = 0.5$ as a function of $x - v_0 T$.

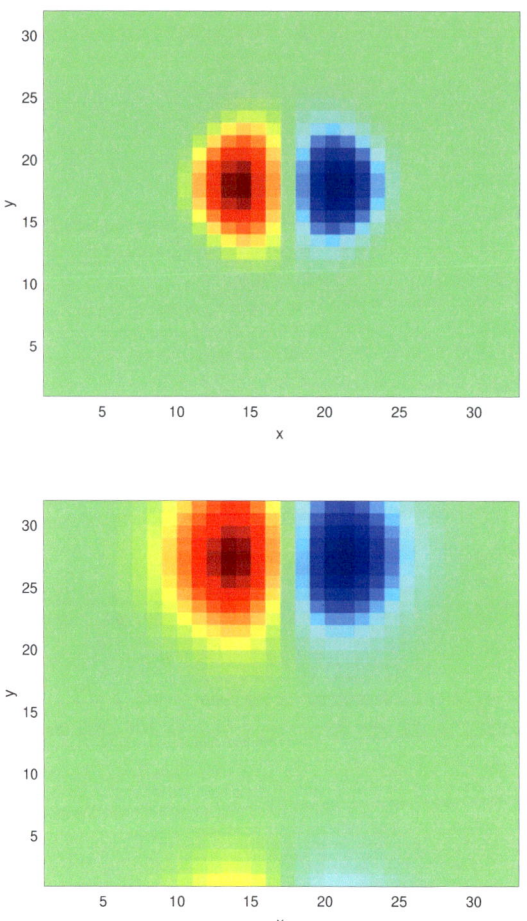

Figure 4.12: Translating vortex: the initial horizontal velocity u at T $= 0$ (top) and at the final time T $= 0.5$ (bottom).

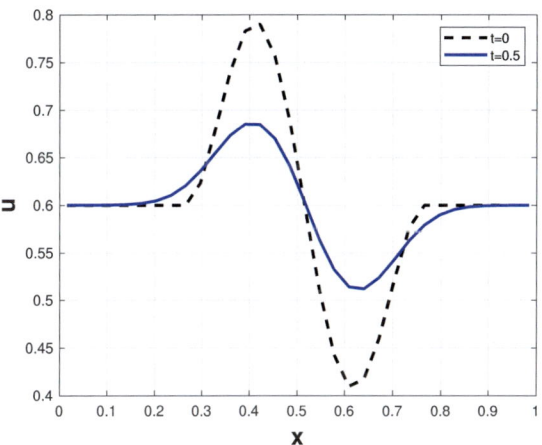

Figure 4.13: Translating vortex: a cross section of the horizontal velocity u at T $= 0$ with $\min_{i,j} u_{i,j+\frac{1}{2}} = 0.4098$ and $\max_{i,j} u_{i,j+\frac{1}{2}} = 0.7902$ and at the final time T $= 0.5$ with $\min_{i,j} u_{i,j+\frac{1}{2}} = 0.52$ and $\max_{i,j} u_{i,j+\frac{1}{2}} = 0.67$ along $x = 0.5 + v_0 T$ as a function of y.

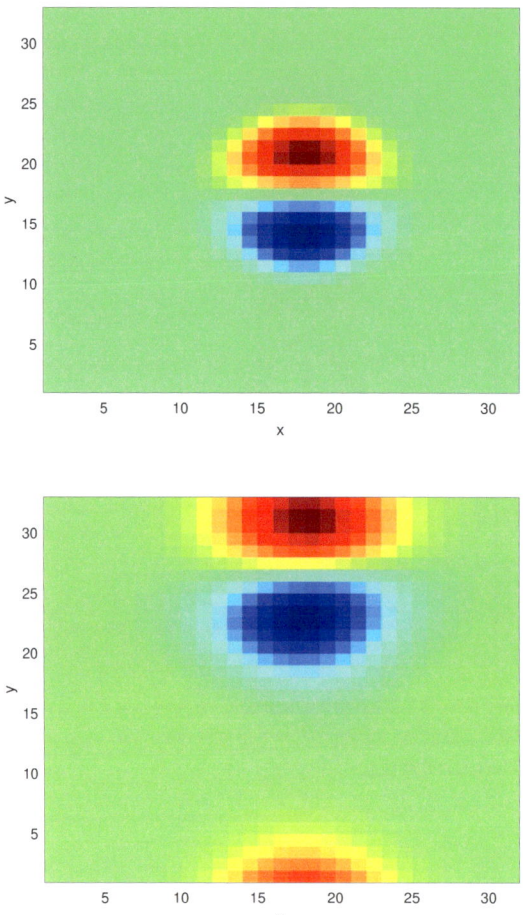

Figure 4.14: Translating vortex: the initial profile of the vertical velocity v at T $= 0$ (top) and at the final time T $= 0.5$ (bottom).

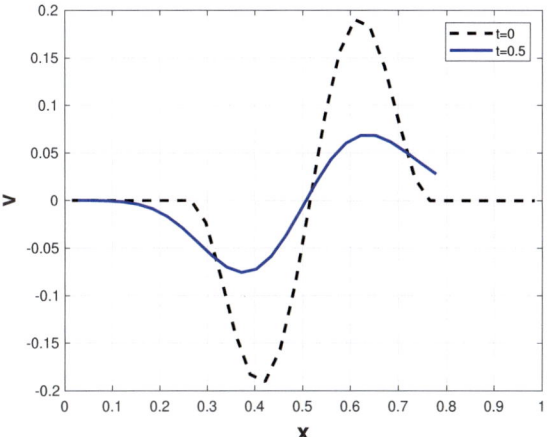

Figure 4.15: Translating vortex: a cross section of the vertical velocity v at $T = 0$ with $\min_{i,j} v_{i+\frac{1}{2},j} = -0.1902$ and $\max_{i,j} v_{i+\frac{1}{2},j} = 0.1902$ and at the final time $T = 0.5$ with $\min_{i,j} v_{i+\frac{1}{2},j} = -0.069$ and $\max_{i,j} v_{i+\frac{1}{2},j} = 0.063$ along $y = 0.5$ as a function of $x - v_0 T$.

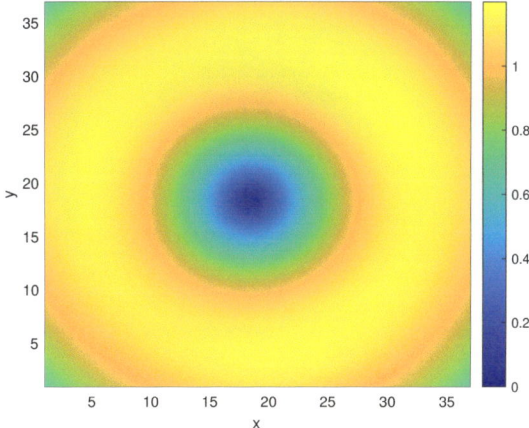

Figure 4.16: Steady vortex: the velocity $q = \sqrt{u^2 + v^2}$ initially on 32 x 32 grid points.

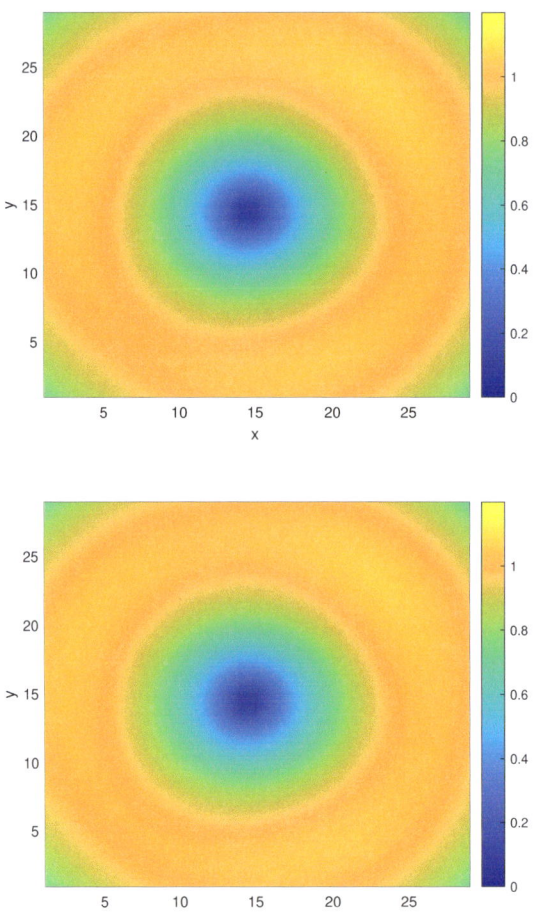

Figure 4.17: Steady vortex: the velocity $q = \sqrt{u^2 + v^2}$ at the final time $T = 1$ for $\varepsilon = 10^{-1}, 10^{-2}$ on 32 x 32 grid points respectively from top left to bottom right.

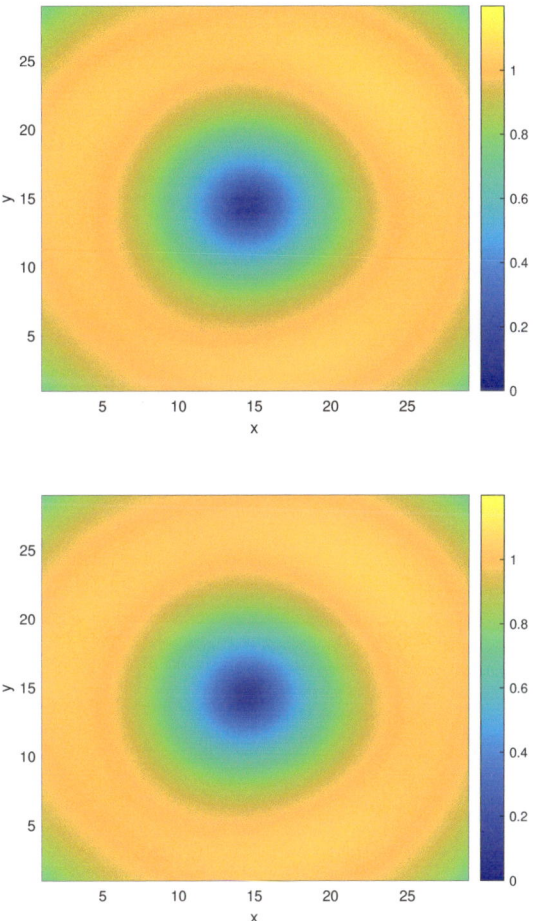

Figure 4.18: Steady vortex: the velocity $q = \sqrt{u^2 + v^2}$ at the final time $T = 1$ for $\varepsilon = 10^{-3}, 10^{-4}$ on 32 x 32 grid points respectively from top left to bottom right.

Chapter 5

Conclusion and Future Work

In this work we investigated the relationship between AP and SP property of a numerical scheme for a parameterized model, such as kinetic equations and low Mach isentropic Euler equations. In other words, we were curious about the long time behavior of a numerical scheme, as well as its behavior as the rescaling parameter approaches zero.

The first aim of this thesis was to develop a well-balanced finite volume central scheme for the system of Euler equations with gravitational source term using the subtraction method, and to extend this well-balancing approach to the system of MHD equations with gravitational source term. Which was succefully accomplished via the subtraction method combined with the CTM in the case of the system of MHD equations.

The second aim was to investigate the SP property of numerical schemes for kinetic models, which became of big interest due to the fact that the Euler equations can be viewed as the limit of the Boltzmann equation. The question was to investigate under which circumstances, AP schemes are SP. The aim was achieved after we introduced a criterion, emphasising that AP schemes with a discretization that linearly depends on the Maxwellian are also SP.

For our third aim, we were interested in projecting the relation between AP and SP schemes for the kinetic models to fluid models. We considered the system of isentropic Euler equations as our first model. In this case, an AP scheme was developed and proven to have the SP property under the condition that the pressure is a function of the density, and the latter is obtained as a solution of an elliptic equation. One interesting extension would be to look at this relation for the full low Mach Euler equations with gravitational source term. Knowing that, in this case, the equation of state is not just the pressure law, and the energy equation is involved. One may also try to find a relation between the low Mach and SP property of the scheme under certain conditions.

To sum up, AP schemes, discretized in a particular way, should be capable of preserving any stationary solutions without any additional treatment. This holds true for kinetic models and for hyperbolic balance laws.

Chapter 6

Appendices

Appendix A. AP property of the UGKS

In this section, we provide a formal derivation of the AP property for the UGKS proposed in (3.17)–(3.18). When ε goes to zero, asymptotic expansions of $A, B,$ and C given in (3.25) read $A = O(\varepsilon), B = \frac{1}{\varepsilon} - \phi(v\sigma_{i+\frac{1}{2}}) + O(\varepsilon), C = -1 + O(\varepsilon)$. The leading order term of (3.18) yields $f_i^{n+1} = \rho_i^{n+1} + O(\varepsilon)$ and we only need to show that (3.17) satisfies the equation for ρ in (3.14), at the discrete level. Suppose that $f_i^n = \rho_i^n + O(\varepsilon)$, then

$$
\begin{cases}
\mathcal{T}^1 f_{i+\frac{1}{2}}^n = \dfrac{1}{2}\left(\rho_i^n + \rho_{i+1}^n\right) + O(\varepsilon), \\[2mm]
\delta^L \mathcal{T}^1 f_{i+\frac{1}{2}}^n = \dfrac{\rho_{i+1}^n - \rho_i^n}{\Delta x} + O(\varepsilon), \\[2mm]
\delta^R \mathcal{T}^1 f_{i+\frac{1}{2}}^n = \dfrac{\rho_{i+1}^n - \rho_i^n}{\Delta x} + O(\varepsilon).
\end{cases}
$$

We deduce that the expansion of $F_{i+\frac{1}{2}}^n$ reads:

$$
F_{i+\frac{1}{2}}^n = -\frac{\rho_i^n + \rho_{i+1}^n}{2|V|}\left(\int_V v\phi(v\sigma_{i+\frac{1}{2}})dv\right) - \frac{\rho_{i+1}^n - \rho_i^n}{3\Delta x} + O(\varepsilon).
$$

Therefore,

$$
\frac{F_{i+\frac{1}{2}}^n - F_{i-\frac{1}{2}}^n}{\Delta x} = -\frac{\rho_{i+1}^n - 2\rho_i^n + \rho_{i-1}^n}{3(\Delta x)^2} + \left(-\left(\frac{1}{|V|}\int_V v\phi(v\sigma_{i+\frac{1}{2}})dv\right)\frac{\rho_i^n + \rho_{i+1}^n}{2}\right.
$$
$$
\left. + \left(\frac{1}{|V|}\int_V v\phi(v\sigma_{i-\frac{1}{2}})dv\right)\frac{\rho_i^n + \rho_{i-1}^n}{2}\right) + O(\varepsilon).
$$

In the limit of $\varepsilon \to 0$, the discretization (3.17) becomes

$$
\frac{\rho_i^{n+1} - \rho_i^n}{\Delta t} = \frac{\rho_{i+1}^n - 2\rho_i^n + \rho_{i-1}^n}{3(\Delta x)^2} + \left(\frac{1}{|V|}\left(\int_V v\phi(v\sigma_{i+\frac{1}{2}})dv\right)\frac{\rho_i^n + \rho_{i+1}^n}{2}\right.
$$
$$
\left. - \frac{1}{|V|}\left(\int_V v\phi(v\sigma_{i-\frac{1}{2}})dv\right)\frac{\rho_i^n + \rho_{i-1}^n}{2}\right),
$$

which is a consistent discretization of the equation for ρ in (3.14). Therefore, the proposed scheme is AP after coupling with the discretization for $S(x,t)$ in (3.15).

Appendix B. AP property of the stationary discretization of the parity equations-based scheme

Consider the behavior of the scheme as $\varepsilon \to 0$ for a stationary discretization of the fully space-time discretized parity equations-based scheme. Equations (3.4) and (3.5) are then,

- Transport step:

$$\begin{cases} v\frac{D^u}{\Delta x}j_i^n = 0 \\ \eta v\frac{D^u}{\Delta x}r_i^n = 0 \end{cases}$$

- Relaxation step:

$$\begin{cases} -\frac{1}{\varepsilon^2}(r_i^n - \rho_{r_i}^n) = 0 \\ -\frac{1}{\varepsilon^2}(j_i^n + (1 - \varepsilon^2\eta)v\frac{D^c}{\Delta x}r_i^n = 0 \end{cases}$$

Consider the relaxation step as $\varepsilon \to 0$,

$$r_i^n = \rho_{r_i}^n \qquad j_i^n = -v\frac{D^c}{\Delta x}\rho_{r_i}^n$$

So,

$$v\frac{D^u}{\Delta x}j_i^n = 0$$

$$\int_0^1 v\frac{D^u}{\Delta x}(-v\frac{D^c}{\Delta x}\rho_{r_i}^n) = 0$$

$$\frac{D^u D^c}{\Delta x^2}\rho_{r_i}^n \int_0^1 v^2 dv = 0$$

$$\frac{1}{3}\frac{D^u D^c}{\Delta x^2}\rho_{r_i}^n = 0$$

which is a consistent discretization of the stationary equation of the diffusion limit. Hence, the discretization of the stationary equation is AP.

Appendix C. AP property of the stationary discretization of UGKS

In this section, we provide a formal derivation of the AP property of the stationary discretization of the UGKS which results from setting $\rho^{n+1} = \rho^n$ and $f^{n+1} = f^n$ in (3.17)–(3.18),

$$\frac{F^n_{i+\frac{1}{2}} - F^n_{i-\frac{1}{2}}}{\Delta x} = 0, \tag{6.1}$$

$$\frac{\Phi^n_{i+\frac{1}{2}} - \Phi^n_{i-\frac{1}{2}}}{\Delta x} = \frac{1}{\varepsilon^2}(\rho^n_i - f^n_i) + \frac{1}{\varepsilon}\left(\frac{1}{|V|}\int_V \phi(v'\sigma_{i+\frac{1}{2}})f^n_i(v')\,dv' - \phi(v\sigma_{i+\frac{1}{2}})f^n_i\right). \tag{6.2}$$

Formulas A, B, and C given in (3.25) are time dependent, but when ε goes to zero, asymptotic expansions of A, B, and C read $A = O(\varepsilon)$, $B = \frac{1}{\varepsilon} - \phi(v\sigma_{i+\frac{1}{2}}) + O(\varepsilon)$, $C = -1 + O(\varepsilon)$ which is time independent. Hence, choosing A, B, and C as in (3.25), for the stationary discretization will not affect the AP proof. We only need to show that (6.1) satisfies the stationary equation of the Keller-Segel equation (3.14) for ρ at the discrete level. Suppose that $f^n_i = \rho^n_i + O(\varepsilon)$, then

$$\begin{cases} \mathcal{T}^1 f^n_{i+\frac{1}{2}} = \frac{1}{2}(\rho^n_i + \rho^n_{i+1}) + O(\varepsilon), \\[2mm] \delta^L \mathcal{T}^1 f^n_{i+\frac{1}{2}} = \frac{\rho^n_{i+1} - \rho^n_i}{\Delta x} + O(\varepsilon), \\[2mm] \delta^R \mathcal{T}^1 f^n_{i+\frac{1}{2}} = \frac{\rho^n_{i+1} - \rho^n_i}{\Delta x} + O(\varepsilon). \end{cases}$$

We deduce that the expansion of $F^n_{i+\frac{1}{2}}$ reads:

$$F^n_{i+\frac{1}{2}} = -\frac{\rho^n_i + \rho^n_{i+1}}{2|V|}\left(\int_V v\phi(v\sigma_{i+\frac{1}{2}})dv\right) - \frac{\rho^n_{i+1} - \rho^n_i}{3\Delta x} + O(\varepsilon).$$

Therefore,

$$\frac{F^n_{i+\frac{1}{2}} - F^n_{i-\frac{1}{2}}}{\Delta x}$$
$$= -\frac{\rho^n_{i+1} - 2\rho^n_i + \rho^n_{i-1}}{3(\Delta x)^2} + \left(-\left(\frac{1}{|V|}\int_V v\phi(v\sigma_{i+\frac{1}{2}})dv\right)\frac{\rho^n_i + \rho^n_{i+1}}{2}\right.$$
$$\left. + \left(\frac{1}{|V|}\int_V v\phi(v\sigma_{i-\frac{1}{2}})dv\right)\frac{\rho^n_i + \rho^n_{i-1}}{2}\right) + O(\varepsilon).$$

In the limit of $\varepsilon \to 0$, the discretization (6.1) becomes

$$\frac{\rho_{i+1}^n - 2\rho_i^n + \rho_{i-1}^n}{3(\Delta x)^2} + \left(\frac{1}{|V|} \left(\int_V v\phi(v\sigma_{i+\frac{1}{2}})dv \right) \frac{\rho_i^n + \rho_{i+1}^n}{2} - \right.$$
$$\left. \frac{1}{|V|} \left(\int_V v\phi(v\sigma_{i-\frac{1}{2}})dv \right) \frac{\rho_i^n + \rho_{i-1}^n}{2} \right) = 0,$$

which is a consistent discretization of the stationary equation for ρ in (3.14). Therefore, the proposed stationary discretization is AP after coupling with the discretization for $S(x,t)$ in (3.15).

Bibliography

[1] M.L. Adams. Discontinuous finite element transport solutions in thick diffusive problems. *Nuclear Science and Engineering*, 137:298–333, 2001.

[2] W. Alt. Biased random walk models for chemotaxis and related diffusion approximations. *Journal of Mathematical Biology*, 9(2):147–177, 1980.

[3] P. Arminjon and R. Touma. Central finite volume methods with constrained transport divergence treatment for ideal mhd. *Journal of Computational Physics*, 204:737–759, 2005.

[4] P. Arminjon and R. Touma. Central finite volume methods with constrained transport divergence treatment for ideal MHD. *Journal of Computational Physics*, 204:737–759, 2005.

[5] P. Arminjon and M-C. Viallon. Généralization du schéma de Nessyahu-Tadmor pour une équation hyperbolique à deux dimensions d'espace. *Comptes Rendus de l'Académie des sciences de Paris*, 320:85–88, 1995.

[6] P. Arminjon and M-C. Viallon. Convergence of a finite volume extension of the Nessyahu-Tadmor scheme on unstructured grid for a two-dimensional linear hyperbolic equation. *SIAM Journal on Numerical Analysis*, 36:738–771, 1999.

[7] P. Arminjon, M-C. Viallon, and A. Madrane. A finite volume extension of the Lax-Friedrichs and Nessyahu-Tadmor schemes for conservation laws on unstructured grids. *International Journal of Computational Fluid Dynamics*, 9(1):1–22, 1998.

[8] J. P. Berberich, P. Chandrashekar, and C. Klingenberg. High order well-balanced finite volume methods for multi-dimensional systems of hyperbolic balance laws.

[9] J. P. Berberich, P. Chandrashekar, and C. Klingenberg. A general well-balanced finite volume scheme for euler equations with gravity. *in: C. Klingenberg, M. Westdickenberg (Eds.), Theory, Numerics and Applications of Hyperbolic Problems I, Springer Proceedings in Mathematics and Statistics*, 236, 2018.

[10] J. P. Berberich, P. Chandrashekar, C. Klingenberg, and F.K. Roepke. Second order finite volume scheme for euler equations with gravity which is well-balanced for general equations of state and grid systems. *Communications in Computational Physics*, 26:599–630, 2019.

[11] T. J. Bogdan and et al. Waves in the magnetized solar atmosphere ii : waves from localized sources in magnetic flux concentrations. *Astrophysics Journal*, 599:626–660, 2003.

[12] N. Botta, S. Langenberg, R. Klein, and S. Lützenkirchen. Well balanced finite volume methods for nearly hydrostatic flows. *Journal of Computational Physics,*, 196:539–565, 2004.

[13] J.U. Brackbill and D.C. Barnes. The effect of nonzero divb on the numerical solution of the magnetohydrodynamic equations. *Journal of Computational Physics*, 35:426–430, 1980.

[14] J.U. Brackbill and D.C. Barnes. The effect of nonzero $\nabla.\mathbf{B}$ on the numerical solution of the magnetohydrodynamic equations. *Journal of Computational Physics*, 201:261–285, 2004.

[15] J.A. Carrillo and B. Yan. An asymptotic preserving scheme for the diffusive limit of kinetic systems for chemotaxis. *Multiscale Modeling & Simulation. A SIAM Interdisciplinary Journal*, 11(1):336–361, 2013.

[16] C. Cercignani. *The Boltzmann equation and its applications*. Springer, 1988.

[17] F. Chalub, P. Markowich, B. Perthame, and C. Schmeiser. Kinetic models for chemotaxis and their drift-diffusion limits. *Monatshefte für Mathematik*, 142(1-2):123–141, 2004.

[18] P. Chandrashekar and C. Klingenberg. A second order well-balanced finite volume scheme for euler equations with gravity. *Journal on Scientific Computing*, 37:B382–B402, 2015.

[19] P. Chandrashekar and M. Zenk. Well-balanced nodal discontinuous galerkin method for euler equations with gravity. *Journal of Scientific Computing*, 1511.08739 v1, 2015.

[20] A. Chertock, S. Cui, A. Kurganov, S.N. özcan, and E. Tadmor. Well-balanced schemes for the euler equations with gravitation: Conservative formulation using global fluxes. 1712.08218v1, 2017.

[21] A. Chertock, A. Kurganov, M.L. Medvidova, and S.N. Özcan. An asymptotic preserving scheme for kinetic chemotaxis models in two space dimensions. *submitted*, 2017.

[22] V. Desveaux, M. Zenk, C. Berthon, and C. Klingenberg. A well-balanced scheme for the euler equation with a gravitational potential. *International Journal for Numerical Methods in Fluids*.

[23] V. Desveaux, M. Zenk, C. Berthon, and C. Klingenberg. A well-balanced scheme for the euler equation with a gravitational potential. *in: Finite Volumes for Complex Applications VII-Methods and Theoretical Aspects, Springer*, pages 217–226, 2014.

[24] V. Desveaux, M. Zenk, C. Berthon, and C. Klingenberg. A well-balanced scheme to capture non-explicit steady states in the euler equations with gravity. *International Journal for Numerical Methods in Fluids*, 81:104–127, 2016.

[25] G. Dimarco and L. Pareschi. Implicit-explicit linear multistep methods for stiff kinetic equations. *SIAM Journal of Numerical Analysis*, 55(2):664–690, 2017.

[26] W.L. Edward, J.E. Morel, and F.M.J. Warren. Asymptotic solutions of numerical transport problems in optically thick, diffusive regimes. *Journal of Computational Physics*, 69(2):283 – 324, 1987.

[27] C. Emako, F. Kanbar, C. Klingenberg, and M. Tang. A criterion for asymptotic preserving schemes of kinetic equations to be uniformly stationary preserving. *Kinetic and Related Models*, 14(5):847–866, 2021.

[28] C.R. Evans and J.F. Hawley. Simulation of magnetohydrodynamic flows: A constrained transport method. *Astrophysics Journal*, 332:659, 1988.

[29] F. Filbet and S. Jin. A class of asymptotic-preserving schemes for kinetic equations and related problems with stiff sources. *Journal of Computational Physics*, 229(20):7625–7648, 2010.

[30] F. Filbet, C. Mouhot, and L. Pareschi. Solving the boltzmann equation in $n \log_2 n$. *SIAM Journal of scientific computation*, 28:1029–1053, 2006.

[31] F.G. Fuchs, A.D. McMurry, S. Mishra, N.H. Risbro, and K. Waagan. High order well-balanced finite volume schemes for simulating wave propagation in stratified magnetic atmospheres. *Journal of Computational Physics*, 229:4033–4058, 2010.

[32] S.K. Godunov. The symmetric form of magnetohydrodynamics equation. *Numerical Methods of Continuum Mechanics. Media 1*, 26, 1972.

[33] L. Gosse. A well-balanced scheme for kinetic models of chemotaxis derived from one-dimensional local forward-backward problems. *Mathematical Biosciences*, 242(2):117–128, 2013.

[34] T. Goudon, J. Liobell, and S. Minjeaud. An asymptotic preserving scheme on staggered grids for the barotropic euler system in low mach regimes. *Numerical Methods for Partial Differential Equations*, pages 1–31, 2020.

[35] L. Grosheintz-Laval and R. Käppeli. High-order well-balanced finite volume schemes for the euler equations with gravitation. *Journal of Computational Physics*, 2019.

[36] A.N. Guarendi and A.J. Chandy. Nonoscillatory central schemes for hyperbolic systems of conservation laws in three-space dimensions. *The Scientific World Journal*, 2013, 2013.

[37] J. Haack, S. Jin, and J-G. Liu. An all-speed asymptotic preserving method for the isentropic euler and navier-stokes equations. *Numerical Methods for Partial Differential Equations*, 2011.

[38] F. H. Harlow and J. E. Welch. Numerical calculation of time-dependent viscous incompressible flow of fluid with free surface. *Physics of Fluids*, 8:2182–2189, 1965.

[39] A. Harten. High resolution schemes for hyperbolic conservation laws. *Journal of Computational Physics*, 49:357–393, 1983.

[40] B. Howard. *E. coli in Motion*. Biological and Medical Physics, Biomedical Engineering. Springer, 2004.

[41] J. Hu and L. Ying. A fast spectral algorithm for the quantum boltzmann collision operator. *Communications in Mathematical Sciences*, 10(3):989–999, 2012.

[42] H.J. Hwang, K. Kang, and A. Stevens. Drift-diffusion limits of kinetic models for chemotaxis: a generalization. *Discrete and Continuous Dynamical Systems. Series B. A Journal Bridging Mathematics and Sciences*, 5(2):319–334, 2005.

[43] C. Ivan and P. Bojan. New non-oscillatory central schemes on unstructured triangulations for hyperbolic systems of conservation laws. *Journal of Computational Physics*, 227(11):5736–5757, May 2008.

[44] G. Jannoun, R. Touma, and F. Brock. Convergence of two-dimensional staggered central schemes on unstructured triangular grids. *Applied Numerical Mathematics*, 92(0):1–20, 6 2015.

[45] G.S. Jiang, D. Levy, C.T. Lin, S. Osher, and E. Tadmor. High-resolution nonoscillatory central schemes with nonstaggered grids for hyperbolic conservation laws. *SIAM Journal on Numerical Analysis*, 35(6):2147–2168, 1998.

[46] G.S. Jiang and E. Tadmor. Nonoscillatory central schemes for multidimensional hyperbolic conservation laws. *SIAM Journal on Scientific Computing*, 19(6):1892–1917, 1998.

[47] S. Jin. Asymptotic preserving (AP) schemes for multiscale kinetic and hyperbolic equations: a review. *Riv. Mat. Univ. Prama*, 3:177–216, 2012.

[48] S. Jin, L. Pareschi, and G. Toscani. Uniformly accurate diffusive relaxation schemes for multiscale transport equations. *SIAM Journal of Numerical Analysis*, 38(3):913–936, 2000.

[49] S. Jin, M. Tang, and H. Han. A uniformly second order numerical method for the one-dimensional discrete-ordinate transport equation and its diffusion limit with interface. *Networks and Heterogeneous Media*, 4(1):35–65, 2009.

[50] F. Kanbar, C. Klingenberg, and M. Tang. Asymptotic and stationary preserving schemes for the isentropic euler equations with gravitational source term. *Manuscript*, 2021.

[51] F. Kanbar, R. Touma, and C. Klingenberg. Well-balanced central schemes for the one and two-dimensional euler systems with gravity. *Applied Numerical Mathematics*, 156:608–626, 2020.

[52] F. Kanbar, R. Touma, and C. Klingenberg. Well-balanced central schemes for the two-dimensional mhd system with gravity. *Manuscript*, 2021.

[53] C. Klingenberg, G. Puppo, and M. Semplice. Arbitrary order finite volume well-balanced schemes for the euler equations with gravity. *SIAM Journal on Scientific Computing*, 41:A695–A721, 2019.

[54] X. Kun. A gas-kinetic BGK scheme for the Navier-stokes equations and its connection with artificial dissipation and Godunov method. *Journal of Computational Physics*, 171(1):289–335, 2001.

[55] X. Kun and H. Juan-Chen. A unified gas-kinetic scheme for continuum and rarefied flows. *Journal of Computational Physics*, 229(20):7747–7764, 2010.

[56] R. Käppeli and S. Mishra. Well-balanced schemes for the euler equations with gravitation. *Journal of computational physics,*, 259:199–219, 2014.

[57] E.W. Larsen and J.E. Morel. Asymptotic solutions of numerical transport problems in optically thick,diffusive regimes ii. *Journal of Computational Physics*, 69:212–236, 1989.

[58] P.D. Lax and B. Wendrof. Systems of conservation laws. *Computations of Pure and Applied Mathematics*, 13:B382–B402, 217-237.

[59] V. Michel-Dansac, C. Berthon, S. Clain, and F. Foucher. A two-dimensional high-order well-balanced scheme for the shallow water equations with topography and manning friction. *hal-02536791*, 2020.

[60] L. Mieussens. On the asymptotic preserving property of the unified gas kinetic scheme for the diffusion limit of linear kinetic models. *Journal of Computational Physics*, 253:138–156, 2013.

[61] H. Nessyahu and E. Tadmor. Non-oscillatory central differencing for hyperbolic conservation laws. *Journal of Computational Physics*, 87(2):408–463, 1990.

[62] H. Othmer, S. Dunbar, and W. Alt. Models of dispersal in biological systems. *Journal of Mathematical Biology*, 26(3):263–298, 1988.

[63] H. Othmer and T. Hillen. The diffusion limit of transport equations. II. Chemotaxis equations. *SIAM Journal on Applied Mathematics*, 62(4):1222–1250 (electronic), 2002.

[64] K.G. Powell. An approximate riemann solver for magnetohydrodynamics (that works in more than one dimension). *ICASE-Report 94-24 (NASA CR-194902) (NASA Langley Research Center, Hampton, VA 23681-0001, 8. April 1994)*.

[65] K.G. Powell. An approximate riemann solver for magnetohydrodynamics (that works in more than one space dimension). *Technical report, 94-24, ICASE, Langley, VA*, 1994.

[66] K.G. Powell, P.L. Roe, T.J. Linde, T.I. Gombosi, and D.L. De zeeuw. A solution adaptive upwind scheme for ideal mhd. *Journal of Computational Physics*, 154(2):284–309, 1999.

[67] K.G. Powell, P.L. Roe, R.S. Myong, T. Gombosi, and D.D. Zeeuw. An upwind scheme for the magnetohydrodynamics. *AIAA Paper 95-1704-CP*, 1995.

[68] G. Puppo, C. Klingenberg, and M. Semplice. Arbitrary order finite volume well-balanced schemes for the euler equations with gravity. *Journal on Scientific Computing*, 2019.

[69] C. S. Rosenthal and et al. Waves in the magnetized solar atmosphere i: Basic processes and internetwork oscillations. *Astrophysics Journal*, 564:508–524, 2002.

[70] J. Saragosti, V. Calvez, N. Bournaveas, A. Buguin, P. Silberzan, and B. Perthame. Mathematical description of bacterial traveling pulses. *PLoS Computational Biology*, 6(8):e1000890, 12, 2010.

[71] J. Saragosti, V. Calvez, N. Bournaveas, B. Perthame, A. Buguin, and P. Silberzan. Directional persistence of chemotactic bacteria in a traveling concentration wave. *Proceedings of the National Academy of Sciences*, 108(39):16235–16240, 2011.

[72] G. Toth. The divb=0 constraint in shock capturing magnetohydrodynamics codes. *Journal of Computational Physics*, 161:605–652, 2000.

[73] R. Touma. Central unstaggered finite volume schemes for hyperbolic systems: Applications to unsteady shallow water equations. *Applied Mathematics and Computation*, 213(1):47–59, 7 2009.

[74] R. Touma. Unstaggered central schemes with constrained transport treatment for ideal and shallow water magnetohydrodynamics. *Applied Numerical Mathematics*, 60(7):752–766, 2010.

[75] R. Touma. Unstaggered central schemes with constrained transport treatment for ideal and shallow water magnetohydrodynamics. *Applied Numerical Mathematics*, 60:752–766, 2010.

[76] R. Touma and P. Arminjon. Central finite volume schemes with constrained transport divergence treatment for three-dimensional ideal mhd. *Journal of Computational Physics*, 212(2):617–636, 2006.

[77] R. Touma and G. Jannoun. Non-oscillatory central schemes on unstructured grids for two-dimensional hyperbolic conservation laws. *Applied Numerical Mathematics*, 62(8):941–955, 2012.

[78] R. Touma and S. Khankan. Well-balanced unstaggered central schemes for one and two-dimensional shallow water equation systems. *Applied Mathematics and Computation*, 218(10):5948–5960, 2012.

[79] R. Touma and C. Klingenberg. Well-balanced central finite volume methods for the ripa system. *Applied Numerical Mathematics*, 97:42–68, 2015.

[80] R. Touma, U. Koley, and C. Klingenberg. Well-balanced unstaggered central schemes for the euler equations with gravitation. *SIAM Journal of Scientific Computing.*, 38(5):773–807, 2016.

[81] D. Varma and P. Chandrashekar. A second-order, discretely well-balanced finite volume scheme for euler equations with gravity. *Computers & Fluids*, 2019.

[82] M.H. Veiga, D.R. Velasco, R. Abgrall, and R. Teyssier. Capturing near-equilibrium solutions: a comparison between high-order discontinuous galerkin methods and well-balanced schemes. *Global Science*, 183.05919 v2, 2018.

[83] W. Kailiang Wu and S. Chi-Wang. Provably positive discontinuous galerkin methods for multidimensional magnetohydrodynamics. *SIAM Journal of Scientific Computing.*, 40:B1302–1329, 2018.

[84] Y. Xing and G. Li. Well-balanced discontinuous galerkin methods for the euler equations under gravitational fields. *Journal of Scientific Computing*, 2016.

[85] Y. Xing and G. Li. Well-balanced discontinuous galerkin methods with hydrostatic reconstruction for the euler equations with gravitation. *Journal of Computational Physics*, 2018.

[86] M. Zenk, A. Thomann, and C. Klingenberg. A second-order positivity-preserving well-balanced finite volume scheme for euler equations with gravity for arbitrary hydrostatic equilibria. 2019.